Charles Davies

**New Elementary Algebra**

Embracing the First Principles of the Science

Charles Davies

**New Elementary Algebra**
*Embracing the First Principles of the Science*

ISBN/EAN: 9783337035457

Printed in Europe, USA, Canada, Australia, Japan

Cover: Foto ©berggeist007 / pixelio.de

More available books at **www.hansebooks.com**

# NEW

# ELEMENTARY ALGEBRA:

EMBRACING

THE FIRST PRINCIPLES OF THE SCIENCE.

BY

## CHARLES DAVIES, LL.D.,

PROFESSOR OF HIGHER MATHEMATICS, COLUMBIA COLLEGE.

———————————

NEW YORK :

BARNES & BURR, PUBLISHERS, 51 & 53 JOHN ST.

CHICAGO: GEORGE SHERWOOD. 118 LAKE ST.
CINCINNATI: RICKEY AND CARROLL.
ST. LOUIS: KEITH AND WOODS.

1864.

*Davies' Course of Mathematics.*

# MATHEMATICAL WORKS,

### IN A SERIES OF THREE PARTS:

## ARITHMETICAL, ACADEMICAL, AND COLLEGIATE

## DAVIES' LOGIC AND UTILITY OF MATHEMATICS.

THIS series, combining all that is most valuable in the various methods of European instruction, improved and matured by the suggestions of more than thirty years' experience, now forms the only complete consecutive course of Mathematics. Its methods, harmonizing as the works of one mind, carry the student onward by the same analogies, and the same laws of association, and are calculated to impart a comprehensive knowledge of the science, combining clearness in the several branches, and unity and proportion in the whole; being the system so long in use at West Point, through which so many men, eminent for their scientific attainments, have passed, and having been adopted, as Text Books, by most of the Colleges in the United States.

### I.—THE ARITHMETICAL COURSE FOR SCHOOLS.

1. PRIMARY ARITHMETIC AND TABLE-BOOK.
2. INTELLECTUAL ARITHMETIC.
3. SCHOOL ARITHMETIC.  (Key separate.)
4. GRAMMAR OF ARITHMETIC.

### II.—THE ACADEMIC COURSE.

1. THE UNIVERSITY ARITHMETIC.  (Key separate.)
2. PRACTICAL MATHEMATICS FOR PRACTICAL MEN.
3. ELEMENTARY ALGEBRA.  (Key separate.)
4. ELEMENTARY GEOMETRY AND TRIGONOMETRY.
5. ELEMENTS OF SURVEYING.

### III.—THE COLLEGIATE COURSE.

1. DAVIES' BOURDON'S ALGEBRA.
2. DAVIES' UNIVERSITY ALGEBRA.
3. DAVIES' LEGENDRE'S GEOMETRY AND TRIGONOMETRY.
4. DAVIES' ANALYTICAL GEOMETRY.
5. DAVIES' DESCRIPTIVE GEOMETRY.
6. DAVIES' SHADES, SHADOWS, AND PERSPECTIVE.
7. DAVIES' DIFFERENTIAL AND INTEGRAL CALCULUS.
8. MATHEMATICAL DICTIONARY, BY DAVIES & PECK.

ENTERED, according to the Act of Congress, in the year one thousand eight hundred and fifty-nine, by CHARLES DAVIES, in the Clerk's Office of the District Court of the United States, for the Southern District of New York.

WILLIAM DENYSE, STEREOTYPER.

# PREFACE.

Algebra naturally follows Arithmetic in a course of scientific studies. The language of figures, and the elementary combinations of numbers, are acquired at an early age. When the pupil passes to a new system, conducted by letters and signs, the change seems abrupt; and he often experiences much difficulty before perceiving that Algebra is but Arithmetic written in a different language.

It is the design of this work to supply a connecting link between Arithmetic and Algebra; to indicate the unity of the methods, and to conduct the pupil from the arithmetical processes to the more abstract methods of analysis, by easy and simple gradations. The work is also introductory to the University Algebra, and to the Algebra of M. Bourdon, which is justly considered, both in this country and in Europe, as the best text-book on the subject, which has yet appeared.

In the Introduction, or Mental Exercises, the language of figures and letters are both employed. Each Lesson is so arranged as to introduce a single principle, not known before, and the whole is so combined as to prepare the pupil, by a thorough system of mental training, for those processes of reasoning which are peculiar to the algebraic analysis.

It is about twenty years since the first publication of the ELEMENTARY ALGEBRA. Within that time, great changes have taken place in the schools of the country. The systems of mathematical instruction have been improved, new methods have been developed, and these require corresponding modifications in the text-books. Those modifications have now been made, and this work will be permanent in its present form.

Many changes have been made in the present edition, at the suggestion of teachers who have used the work, and favored me with their opinions, both of its defects and merits. I take this opportunity of thanking them for the valuable aid they have rendered me. The criticisms of those engaged in the daily business of teaching are invaluable to an author; and I shall feel myself under special obligation to all who will be at the trouble to communicate. to me, at any time, such changes, either in methods or language, as their experience may point out. It is only through the cordial co-operation of teachers and authors—by joint labors and mutual efforts—that the text-books of the country can be brought to any reasonable degree of perfection.

COLUMBIA COLLEGE, NEW YORK, *March*, 1859.

# CONTENTS.

## CHAPTER I.

### DEFINITIONS AND EXPLANATORY SIGNS.

## CHAPTER II.

### FUNDAMENTAL OPERATIONS.

## CHAPTER III.

### USEFUL FORMULAS. FACTORING, ETC.

## CHAPTER IV

### FRACTIONS.

v

## CHAPTER V.

### EQUATIONS OF THE FIRST DEGREE.

## CHAPTER VI.

### FORMATION OF POWERS.

## CHAPTER VII.

### SQUARE ROOT. RADICALS OF THE SECOND DEGREE.

# SUGGESTIONS TO TEACHERS.

1. THE Introduction is designed as a mental exercise. If thoroughly taught, it will train and prepare the mind of the pupil for those higher processes of reasoning, which it is the peculiar province of the algebraic analysis to develop.

2. The statement of each question should be made, and every step in the solution gone through with, without the aid of a slate or black-board; though perhaps, in the beginning, some aid may be necessary to those unaccustomed to such exercises.

3. Great care must be taken to have every principle on which the statement depends, carefully analyzed; and equal care is necessary to have every step in the solution distinctly explained.

4. The reasoning process is the logical connection of distinct apprehensions, and the deduction of the consequences which follow from such a connection. Hence, the basis of all reasoning must lie in distinct elementary ideas.

5. Therefore, to teach one thing at a time—to teach that thing well—to explain its connections with other things, and the consequences which follow from such connections, would seem to embrace the whole art of instruction.

# ELEMENTARY ALGEBRA.

---

## MENTAL EXERCISES.

### LESSON I.

1. JOHN and Charles have the same number of apples; both together have twelve: how many has each?

ANALYSIS.—Let $x$ denote the number which John has; then, since they have an equal number, $x$ will also denote the number which Charles has, and twice $x$, or $2x$, will denote the number which both have, which is 12. If twice $x$ is equal to 12, $x$ will be equal to 12 divided by 2, which is 6; therefore, each has 6 apples.

WRITTEN.

Let $x$ denote the number of apples which John has; then,

$$x + x = 2x = 12; \text{ hence, } x = \frac{12}{2} = 6.$$

NOTE.—When $x$ is written with the sign $+$ before it, it is read *plus* $x$: and the line above, is read, $x$ *plus* $x$ *equals* 12.

NOTE.—When $x$ is written by itself, it is read one $x$, and is the same as, $1x$;

| | | | | | |
|---|---|---|---|---|---|
| $x$ or $1x$, | means | once | $x$, | or one | $x$, |
| $2x$, | " | twice | $x$, | or two | $x$, |
| $3x$, | " | three times | $x$, | or three | $x$, |
| $4x$, | " | four times | $x$, | or four | $x$, |
| &c., | | &c., | | &c. | |

2. What is $x + x$ equal to?

3. What is $x + 2x$ equal to?

4. What is $x + 2x + x$ equal to?

5. What is $x + 5x + x$ equal to?

6. What is $x + 2x + 3x$ equal to?

7. James and John together have twenty-four peaches, and one has as many as the other: how many has each?

ANALYSIS.—Let $x$ denote the number which James has; then, since they have an equal number, $x$ will also denote the number which John has, and twice $x$ will denote the number which both have, which is 24. If twice $x$ is equal to 24, $x$ will be equal to 24 divided by 2, which is 12; therefore, each has 12 peaches.

WRITTEN.

Let $x$ denote the number of peaches which James has; then,

$$x + x = 2x = 24; \text{ hence, } x = \frac{24}{2} = 12.$$

VERIFICATION.

*A Verification* is the operation of proving that the number found will satisfy the conditions of the question. Thus,

James' apples.   John's apples.

12       + 12     = 24.

NOTE.—Let the following questions be *analyzed*, *written*, and *verified*, in *exactly* the *same manner* as the above.

8. William and John together have 36 pears, and one has as many as the other : how many has each?

9. What number added to itself will make 20?

10. James and John are of the same age, and the sum of their ages is 32: what is the age of each?

11. Lucy and Ann are twins, and the sum of their ages is 16: what is the age of each?

12. What number is that which added to itself will make 30?

13. What number is that which added to itself will make 50?

14. Each of two boys received an equal sum of money at Christmas, and together they received 60 cents: how much had each?

15. What number added to itself will make 100?

16. John has as many pears as William; together they have 72: how many has each?

17. What number added to itself will give a sum equal to 46?

18. Lucy and Ann have each a rose bush with the same number of buds on each; the buds on both number 46: how many on each?

———

## LESSON II.

1. John and Charles together have 12 apples, and Charles has twice as many as John: how many has each?

ANALYSIS.—Let $x$ denote the number of apples which John has; then, since Charles has twice as many, $2x$ will denote his share, and $x + 2x$, or $3x$, will denote the number which they both have, which is 12. If $3x$ is equal to 12, $x$ will be equal to 12 divided by 3, which is 4; therefore, John has 4 apples, and Charles, having twice as many, has 8.

WRITTEN.

Let $x$ denote the number of apples John has; then,

$2x$ will denote the number of apples Charles has; and

$x + 2x = 3x = 12$, the number both have; then,

$x = \dfrac{12}{3} = 4$, the number John has; and,

$2x = 2 \times 4 = 8$, the number Charles has.

VERIFICATION.

$4 + 8 = 12$, the number both have.

2. William and John together have 48 quills, and William has twice as many as John: how many has each?

3. What number is that which added to twice itself, will give a number equal to 60?

4. Charles' marbles added to John's make 3 times as many as John has; together they have 51: how many has each?

ANALYSIS.—Since Charles' marbles added to John's make three times as many as Charles has, Charles must have one third, and John two thirds of the whole.

Let $x$ denote the number which Charles has; then $2x$ will denote the number which John has, and $x + 2x$, or $3x$, will denote what they both have, which is 51. Then, if $3x$ is equal to 51, $x$ will be equal to 51 divided by 3, which is 17. Therefore, Charles has 17 marbles, and John, having twice as many, has 34.

WRITTEN.

Let $x$ denote the number of Charles' marbles; then,

$2x$ will denote the number of John's marbles; and

$3x = 51$, the number of both; then,

$x = \dfrac{51}{3} = 17$, Charles' marbles; and

$17 \times 2 = 34$, John's marbles.

5. What number added to twice itself will make 75 ?

6. What number added to twice itself will make 57 ?

7. What number added to twice itself will make 39 ?

8. What number added to twice itself will give 90 ?

9. John walks a certain distance on Tuesday, twice as far on Wednesday, and in the two days he walks 27 miles · how far did he walk each day ?

10. Jane's bush has twice as many roses as Nancy's: and on both bushes there are 36: how many on each?

11. Samuel and James bought a ball for 48 cents; Samuel paid twice as much as James: what did each pay?

12. Divide 48 into two such parts that one shall be double the other.

13. Divide 66 into two such parts that one shall be double the other.

14. The sum of three equal numbers is 12: what are the numbers?

ANALYSIS.—Let $x$ denote one of the numbers; then, since the numbers are equal, $x$ will also denote each of the others, and $x$ plus $x$ plus $x$, or $3x$ will denote their sum, which is 12. Then, if $3x$ is equal to 12, $x$ will be equal to 12 divided by 3, which is 4: therefore, the numbers are 4, 4, and 4.

WRITTEN.

Let $x$ denote one of the equal numbers; then,

$$x + x + x = 3x = 12; \text{ and}$$
$$x = \frac{12}{3} = 4.$$

VERIFICATION.

$$4 + 4 + 4 = 12.$$

15. The sum of three equal numbers is 24: what are the numbers?

16. The sum of three equal numbers is 36: what are the numbers?

17. The sum of three equal numbers is 54 : what are the numbers?

---

## LESSON III.

1. What number is that which added to three times itself will make 48?

ANALYSIS.—Let $x$ denote the number; then, $3x$ will denote three times the number, and $x$ plus $3x$, or $4x$, will denote the sum, which is 48. If $4x$ is equal to 48, $x$ will be equal to 48 divided by 4, which is 12; there-fore, 12 is the required number.

### WRITTEN.

Let $x$ denote the number; then,

$3x$ = three times the number; and

$x + 3x = 4x = 48$, the sum: then,

$$x = \frac{48}{4} = 12, \text{ the required number.}$$

### VERIFICATION.

$$12 \times 3 + 12 = 12 + 36 = 48.$$

NOTE.—All *similar* questions are solved by the same form of analysis.

2. What number added to 4 times itself will give 40?

3. What number added to 5 times itself will give 42?

4. What number added to 6 times itself will give 63?

5. What number added to 7 times itself will give 84?

6. What number added to 8 times itself will give 81?

7. What number added to 9 times itself will give 100?

8. James and John together have 24 quills, and John has three times as many as James: how many has each?

9. William and Charles have 64 marbles, and Charles has 7 times as many as William: how many has each?

10. James and John travel 96 miles, and James travels 11 times as far as John : how far does each travel ?

11. The sum of the ages of a father and son is 84 years; and the father is 3 times as old as the son : what is the age of each ?

12. There are two numbers of which the greater is 7 times the less, and their sum is 72 : what are the numbers?

13. The sum of four equal numbers is 64 : what are the numbers ?

14. The sum of six equal numbers is 54 : what are the numbers ?

15. James has 24 marbles ; he loses a certain number, and then gives away 7 times as many as he loses which takes all he has : how many did he give away? Verify.

16. William has 36 cents, and divides them between his two brothers, James and Charles, giving one, eight times as many as the other : how many does he give to each?

17. What is the sum of $x$ and $3x$? Of $x$ and $7x$? Of $x$ and $5x$? Of $x$ and $12x$?

---

## LESSON IV.

1. If 1 apple costs 1 cent, what will a number of apples denoted by $x$ cost?

ANALYSIS.—Since one apple costs 1 cent, and since $x$ denotes *any number* of apples, the cost of $x$ apples will be as many cents as there are apples : that is, $x$ cents.

2. If 1 apple costs 2 cents, what will $x$ apples cost?

ANALYSIS.—Since one apple costs 2 cents, and since $x$ denotes the number of apples, the cost will be twice as many cents as there are apples : that is $2x$ cents.

3. If 1 apple costs 3 cents, what will $x$ apples cost ?

4. If 1 lemon costs 4 cents, what will $x$ lemons cost ?

5. If 1 orange costs 6 cents, what will $x$ oranges cost?

6. Charles bought a certain number of lemons at 2 cents apiece, and as many oranges at 3 cents apiece, and paid in all 20 cents: how many did he buy of each?

ANALYSIS.—Let $x$ denote the number of lemons; then, since he bought as many oranges as lemons, it will also denote the number of oranges. Since the lemons were 2 cents apiece, $2x$ will denote the cost of the lemons; and since the oranges were 3 cents apiece, $3x$ will denote the cost of the oranges; and $2x + 3x$, or $5x$, will denote the cost of both, which is 20 cents. Now, since $5x$ cents are equal to 20 cents, $x$ will be equal to 20 cents divided by 5 cents, which is 4: hence, he bought 4 of each.

### WRITTEN.

Let $x$ denote the number of lemons, or oranges; then,

$2x =$ the cost of the lemons; and

$3x =$ the cost of the oranges; hence,

$2x + 3x = 5x = 20$ cents $=$ the cost of lemons and oranges; hence,

$$x = \frac{20 \text{ cents}}{5 \text{ cents}} = 4, \text{ the number of each.}$$

### VERIFICATION.

4 lemons at 2 cents each, give, $4 \times 2 = 8$ cents.

4 oranges at 3 cents each, " $4 \times 3 = 12$ cents.

Hence, they both cost, 8 cents $+12$ cents $= 20$ cents.

7. A farmer bought a certain number of sheep at 4 dollars apiece, and an equal number of lambs at 1 dollar apiece, and the whole cost 60 dollars: how many did he buy of each?

8. Charles bought a certain number of apples at 1 cent apiece, and an equal number of oranges at 4 cents apiece, and paid 50 cents in all: how many did he buy of each?

9. James bought an equal number of apples, pears, and lemons; he paid 1 cent apiece for the apples, 2 cents apiece for the pears, and 3 cents apiece for the lemons, and paid 72 cents in all: how many did he buy of each? Verify.

10. A farmer bought an equal number of sheep, hogs, and calves, for which he paid 108 dollars; he paid 3 dollars apiece for the sheep, 5 dollars apiece for the hogs, and 4 dollars apiece for the calves: how many did he buy of each?

11. A farmer sold an equal number of ducks, geese, and turkeys, for which he received 90 shillings. The ducks brought him 3 shillings apiece, the geese 5, and the turkeys 7: how many did he sell of each sort?

12. A tailor bought, for one hundred dollars, two pieces of cloth, each of which contained an equal number of yards. For one piece he paid 3 dollars a yard, and for the other 2 dollars a yard: how many yards in each piece?

13. The sum of three numbers is 28; the second is twice the first, and the third twice the second: what are the numbers? Verify.

14. The sum of three numbers is 64; the second is 3 times the first, and the third 4 times the second: what are the numbers?

## LESSON V.

1. If 1 yard of cloth costs $x$ dollars, what will 2 yards cost?

ANALYSIS.—Two yards of cloth will cost twice as much as one yard. Therefore, if 1 yard of cloth costs $x$ dollars, 2 yards will cost twice $x$ dollars, or $2x$ dollars.

2. If 1 yard of cloth costs $x$ dollars, what will 3 yards cost? Why?

3. If 1 orange costs $x$ cents, what will 7 oranges cost?
Why? 8 oranges?

4. Charles bought 3 lemons and 4 oranges, for which he
paid 22 cents. He paid twice as much for an orange as for
a lemon : what was the price of each?

ANALYSIS.—Let $x$ denote the price of a lemon; then, $2x$
will denote the price of an orange; $3x$ will denote the cost
of 3 lemons, and $8x$ the cost of 4 oranges; hence, $3x$ plus
$8x$, or $11x$, will denote the cost of the lemons and oranges,
which is 22 cents. If $11x$ is equal to 22 cents, $x$ is equal to
22 cents divided by 11, which is 2 cents: therefore, the
price of 1 lemon is 2 cents, and that of 1 orange 4 cents.

### WRITTEN.

Let $x$ denote the price of 1 lemon; then,

$2x = $ " 1 orange; and,

$3x + 8x = 11x = $ 22 cts., the cost of lemons and oranges;

hence, $x = \dfrac{22 \text{ cts.}}{11} = $ 2 cts., the price of 1 lemon;

and, $2 \times 2 = $ 4 cts., the price of 1 orange.

### VERIFICATION.

$3 \times 2 = $ 6 cents, cost of lemons,

$4 \times 4 = $ 16 cents, cost of oranges.

22 cents, total cost.

5. James bought 8 apples and 3 oranges, for which he
paid 20 cents. He paid as much for 1 orange as for 4 apples:
what did he pay for one of each?

6. A farmer bought 3 calves and 7 pigs, for which he paid
19 dollars. He paid four times as much for a calf as for a
pig : what was the price of each?

7. James bought an apple, a peach, and a pear, for which
he paid 6 cents. He paid twice as much for the peach as for

the apple, and three times as much for the pear as for the apple: what was the price of each?

8. William bought an apple, a lemon, and an orange, for which he paid 24 cents. He paid twice as much for the lemon as for the apple, and 3 times as much for the orange as for the apple: what was the price of each?

9. A farmer sold 4 calves and 5 cows, for which he received 120 dollars. He received as much for 1 cow as for 4 calves: what was the price of each?

10. Lucy bought 3 pears and 5 oranges, for which she paid 26 cents, giving twice as much for each orange as for each pear: what was the price of each?

11. Ann bought 2 skeins of silk, 3 pieces of tape, and a penknife, for which she paid 80 cents. She paid the same for the silk as for the tape, and as much for the penknife as for both: what was the cost of each?

12. James, John, and Charles are to divide 56 cents among them, so that John shall have twice as many as James, and Charles twice as many as John: what is the share of each?

13. Put 54 apples into three baskets, so that the second shall contain twice as many as the first, and the third as many as the first and second: how many will there be in each.

14. Divide 60 into four such parts that the second shall be double the first, the third double the second, and the fourth double the third: what are the numbers?

## LESSON VI.

1. If $2x + x$ is equal to $3x$, what is $3x - x$ equal to? Written, $$3x - x = 2x.$$

2. What is $4x - x$ equal to? Written, $$4x - x = 3x.$$

3. What is $8x$ minus $6x$ equal to? Written,

$$8x - 6x = 2x.$$

4. What is $12x - 9x$ equal to?                *Ans.* 3x.

5. What is $15x - 7x$ equal to?

6. What is $17x - 13x$ equal to?               *Ans.* 4x.

7. Two men, who are 30 miles apart, travel towards each other; one at the rate of 2 miles an hour, and the other at the rate of 3 miles an hour: how long before they will meet?

ANALYSIS.—Let $x$ denote the number of hours. Then, since the time, multiplied by the rate, will give the distance, $2x$ will denote the distance traveled by the first, and $3x$ the distance traveled by the second. But the sum of the distances is 30 miles; hence,

$$2x + 3x = 5x = 30 \text{ miles};$$

and if $5x$ is equal to 30, $x$ is equal to 30 divided by 5, which is 6: hence, they will meet in 6 hours.

### WRITTEN.

Let $x$ denote the time in hours; then,

$$2x = \text{ the distance traveled by the 1st; and}$$
$$3x = \qquad \text{``} \qquad \text{``} \qquad \text{2d.}$$

By the conditions,

$$2x + 3x = 5x = 30 \text{ miles, the distance apart;}$$

hence,                $$x = \frac{30}{5} = 6 \text{ hours.}$$

### VERIFICATION.

$2 \times 6 = 12$ miles, distance traveled by the first.

$3 \times 6 = 18$ miles, distance traveled by the second.

30 miles, whole distance.

8. Two persons are 10 miles apart, and are traveling in the same direction; the first at the rate of 3 miles an hour, and the second at the rate of 5 miles: how long, before the second will overtake the first?

ANALYSIS.—Let $x$ denote the time, in hours. Then, $3x$ will denote the distance traveled by the first in $x$ hours; and $5x$ the distance traveled by the second. But when the second overtakes the first, he will have traveled 10 miles more than the first: hence,

$$5x - 3x = 2x = 10;$$

if $2x$ is equal to 10, $x$ is equal to 5 : hence, the second will overtake the first in 5 hours.

### WRITTEN.

Let $x$ denote the time, in hours: then,

$3x =$ the distance traveled by the 1st;

and, $5x =$      "      "      2d;

and, $5x - 3x = 2x = 10$ hours;

or,          $x = \dfrac{10}{2} = 5$ hours.

### VERIFICATION.

$3 \times 5 = 15$ miles, distance traveled by 1st.

$5 \times 5 = 25$ miles,    "    "    2d.

$25 - 15 = 10$ miles, distance apart.

9. A cistern, holding 100 hogsheads, is filled by two pipes; one discharges 8 hogsheads a minute, and the other 12 : in what time will they fill the cistern?

10. A cistern, holding 120 hogsheads, is filled by 3 pipes; the first discharges 4 hogsheads in a minute, the second 7, and the third 1 : in what time will they fill the cistern?

11. A cistern which holds 90 hogsheads, is filled by a pipe which discharges 10 hogsheads a minute; but there is a waste pipe which loses 4 hogsheads a minute : how long will it take to fill the cistern?

12. Two pieces of cloth contain each an equal number of yards; the first cost 3 dollars a yard, and the second 5, and both pieces cost 96 dollars : how many yards in each?

13. Two pieces of cloth contain each an equal number of yards; the first cost 7 dollars a yard, and the second 5; the first

cost 60 dollars more than the second: how many yards in each piece?

14. John bought an equal number of oranges and lemons the oranges cost him 5 cents apiece, and the lemons 3; and he paid 56 cents for the whole: how many did he buy of each kind?

15. Charles bought an equal number of oranges and lemons; the oranges cost him 5 cents apiece, and the lemons 3; he paid 14 cents more for the oranges than for the lemons: how many did he buy of each?

16. Two men work the same number of days, the one receives 1 dollar a day, and the other two: at the end of the time they receive 54 dollars: how long did they work?

## LESSON VII.

1. John and Charles together have 25 cents, and Charles has 5 more than John: how many has each?

ANALYSIS.—Let $x$ denote the number which John has; then, $x + 5$ will denote the number which Charles has, and $x + x + 5$, or $2x + 5$, will be equal to 25, the number they both have. Since $2x + 5$ equals 25, $2x$ will be equal to 25 minus 5, or 20, and $x$ will be equal to 20 divided by 2, or 10: therefore, John has 10 cents, and Charles 15.

WRITTEN.

Let $x$ denote the number of John's cents; then,

$$x + 5 = \quad \text{``} \quad \text{Charles' cents; and,}$$
$$x + x + 5 = 25, \text{ the number they both have; or,}$$
$$2x + 5 = 25; \quad \text{and,}$$
$$2x = 25 - 5 = 20; \quad \text{hence,}$$
$$x = \frac{20}{2} = 10, \text{ John's number; and,}$$
$$10 + 5 = 15, \text{ Charles' number.}$$

VERIFICATION.

| John's. | Charles'. | | |
|---|---|---|---|
| 10 | + 15 | = 25, | the sum. |

| Charles'. | John's. | | |
|---|---|---|---|
| 15 | — 10 | = 5, | the difference. |

2. James and John have 30 marbles, and John has 4 more tnan James: how many has each?

3. William bought 60 oranges and lemons; there were 20 more lemons than oranges: how many were there of each sort?

4. A farmer has 20 more cows than calves; in all he has 36: how many of each sort?

5. Lucy has 28 pieces of money in her purse, composed of cents and dimes; the cents exceed the dimes in number by 16: how many are there of each sort?

6. What number added to itself, and to 9, will make 29?

7. What number added to twice itself, and to 4, will make 25?

8. What number added to three times itself, and to 12, will make 60?

9. John has five times as many marbles as Charles, and what they both have, added to 14, makes 44: how many has each?

10. There are three numbers, of which the second is twice the first, and the third twice the second, and when 9 is added to the sum, the result is 30: what are the numbers?

11. Divide 13 into two such parts that the second shall be two more than double the first: what are the parts?

12. Divide 50 into three such parts that the second shall be twice the first, and the third exceed six times the first by 4: what are the parts?

13. Charles has twice as many cents as James, and John

has twice as many as Charles; if 7 be added to what they all have, the sum will be 28: how many has each?

14. Divide 15 into three such parts that the second shall be 3 times the first, the third twice the second, and 5 over: what are the numbers?

15. An orchard contains three kinds of trees, apples, pears, and cherries; there are 4 times as many pears as apples, twice as many cherries as .pears, and if 14 be added, the number will be 40; how many are there of each?

---

## LESSON VIII.

1. John after giving away 5 marbles, had 12 left: how many had he at first?

ANALYSIS.—Let $x$ denote the number; then, $x$ minus 5 will denote what he had left, which was equal to 12. Since $x$ diminished by 5 is equal to 12, $x$ will be equal to 12, increased by 5; that is, to 17: therefore, he had 17 marbles.

### WRITTEN.

Let $x$ denote the number he had at first; then,
$$x - 5 = 12,$$ what he had left; and
$$x = 12 + 5 = 17,$$ what he first had.

### VERIFICATION.

$$17 - 5 = 12,$$ what were left.

2. Charles lost 6 marbles and has 9 left: how many had he at first?

3. William gave 15 cents to John, and had 9 left: how many had he at first?

4. Ann plucked 8 buds from her rose bush, and there were 10 left: how many were there at first?

5. William took 27 cents from his purse, and there were 13 left: how many were there at first?

6. The sum of two numbers is 14, and their difference is 2: what are the numbers?

ANALYSIS.—The difference of two numbers, added to the less, will give the greater. Let $x$ denote the *less* number; then, $x + 2$, will denote the greater, and $x + x + 2$, will denote their sum, which is 14. Then, $2x + 2$ equals 14; and $2x$ equals 14 minus 2, or 12: hence, $x$ equals 12 divided by 2, or 6: hence, the numbers are 6 and 8.

### VERIFICATION.

$6 + 8 = 14$, their sum; and
$8 - 6 = 2$, their difference.

7. The sum of two numbers is 18, and their difference 6: what are the numbers?

8. James and John have 26 marbles, and James has 4 more than John: how many has each?

9. Jane and Lucy have 16 books, and Lucy has 8 more than Jane: how many has each?

10. William bought an equal number of oranges and lemons; Charles took 5 lemons, after which William had but 25 of both sorts: how many did he buy of each?

11. Mary has an equal number of roses on each of two bushes; if she takes 4 from one bush, there will remain 24 on both: how many on each at first?

12. The sum of two numbers is 20, and their difference is 6: what are the numbers?

ANALYSIS.—If $x$ denotes the *greater* number, $x - 6$ will denote the less, and $x + x - 6$ will be equal to 20; hence, $2x$ equals $20 + 6$, or 26, and $x$ equals 26 divided by 2, equals 13; hence the numbers are 13 and 7.

2

Let $x$ denote the greater; then,

$x - 6 =$ the less; and

$x + x - 6 = 20$, their sum; hence,

$2x = 20 + 6 = 26$; or,

$x = \dfrac{26}{2} = 13$; and $13 - 6 = 7$.

VERIFICATION.

$13 + 7 = 20$; and, $13 - 7 = 6$.

13. The sum of the ages of a father and son is 60 years, and their difference is just half that number: what are their ages?

14. The sum of two numbers is 23, and the larger lacks 1 of being 7 times the smaller: what are the numbers?

15. The sum of two numbers is 50; the larger is equal to 10 times the less, minus 5: what are the numbers?

16. John has a certain number of oranges, and Charles has four times as many, less seven; together they have 53: how many has each?

17. An orchard contains a certain number of apple trees, and three times as many cherry trees, less 6; the whole number is 30: how many of each sort?

---

## LESSON IX.

1. If $x$ denotes any number, and 1 be added to it, what will denote the sum?        *Ans.* $x + 1$.

2. If 2 be added to $x$, what will denote the sum? If 3 be added, what? If 4 be added? &c.

If to John's marbles, one marble be added, twice his number will be equal to 10: how many had he?

ANALYSIS.—Let $x$ denote the number; then, $x + 1$ will denote the number after 1 is added, and twice this number, or $2x + 2$, will be equal to 10. If $2x + 2$ is equal to 10, $2x$ will be equal to 10 minus 2, or 8; or $x$ will be equal to 4.

### WRITTEN.

Let $x$ denote the number of John's marbles; then,

$$x + 1 = \text{the number, after 1 is added; and}$$

$$2(x + 1) = 2x + 2 = 10; \text{ hence,}$$

$$2x = 10 - 2; \text{ or } x = \frac{8}{2} = 4.$$

### VERIFICATION.

$$2(4 + 1) = 2 \times 5 = 10.$$

4. Write $x + 2$ multiplied by 3.      *Ans.* $3(x + 2)$.
What is the product?      *Ans.* $3x + 6$.

5. Write $x + 4$ multiplied by 5.      *Ans.* $5(x + 4)$.
What is the product?      *Ans.* $5x + 20$.

6. Write $x + 3$ multiplied by 4.      *Ans.* $4(x + 3)$.
What is the product?      *Ans.* $4x + 12$.

7. Lucy has a certain number of books; her father gives her two more, when twice her number is equal to 14: how many has she?

8. Jane has a certain number of roses in blossom; two more bloom, and then 3 times the number is equal to 15: how many were in blossom at first?

9. Jane has a certain number of handkerchiefs, and buys 4 more, when 5 times her number is equal to 45: how many had she at first?

10. John has 1 apple more than Charles, and 3 times John's, added to what Charles has, make 15: how many has each?

ANALYSIS.—Let $x$ denote Charles' apples; then $x + 1$ will denote John's; and $x + 1$ multiplied by 3, added to $x$, or $3x + 3 + x$, will be equal to 15, what they both had; hence, $4x + 3$ equals 15; and $4x$ equals 15 minus 3, or 12; and $x = 4$. Write, and verify.

11. James has two marbles more than William, and twice his marbles plus twice William's are equal to 16: how many has each?

12. Divide 20 into two such parts that one part shall exceed the other by 4.

13. A fruit-basket contains apples, pears, and peaches; there are 2 more pears than apples, and twice as many peaches as pears; there are 22 in all: how many of each sort?

14. What is the sum of $x + 3x + 2(x + 1)$?

15. What is the sum of $2(x + 1) + 1(x + 1) + x$?

16. What is the sum of $x + 5(x + 8)$?

17. The sum of two numbers is 11, and the second is equal to twice the first plus 4: what are the numbers?

18. John bought 3 apples, 3 lemons, and 3 oranges, for which he paid 27 cents; he paid 1 cent more for a lemon than for an apple, and 1 cent more for an orange than for a lemon: what did he pay for each?

19. Lucy, Mary, and Ann, have 15 cents; Mary has 1 more than Lucy, and Ann twice as many as Mary?

---

## LESSON X.

1. If $x$ denote any number, and 1 be subtracted from it, what will denote the difference?　　　　*Ans.* $x - 1$.

If 2 be subtracted, what will denote the difference? If 3 be subtracted? 4? &c.

2. John has a certain number of marbles; if 1 be taken away, twice the remainder will be equal to 12: how many has he?

ANALYSIS.—Let $x$ denote the number; then, $x - 1$ will denote the number after 1 is taken away; and twice this number, or $2(x - 1) = 2x - 2$, will be equal to 12. If $2x$

diminished by 2 is equal to 12, $2x$ is equal to 12 plus 2, or 14; hence, $x$ equals 14 divided by 2, or 7.

WRITTEN.

Let $x$ denote the number; then,

$x - 1 =$ the number which remained, and

$2(x - 1) = 2x - 2 = 12$; hence,

$2x = 12 + 2$, or 14; and $x = \dfrac{14}{2} = 7$.

VERIFICATION.

$2(7 - 1) = 14 - 2 = 12$; also, $2(7 - 1) = 2 \times 6 = 12$

3. Write 3 times $x - 1$.      *Ans.* $3(x - 1)$.

What is the product equal to?      *Ans.* $3x - 3$.

4. Write 4 times $x - 2$.      *Ans.* $4(x - 2)$.

What is the product equal to?      *Ans.* $4x - 8$.

5. Write 5 times $x - 5$.      *Ans.* $5(x - 5)$.

What is the product equal to?      *Ans.* $5x - 25$.

6. If $x$ denotes a certain number, will $x - 1$ denote a greater or less number? how much less?

7. If $x - 1$ is equal to 4, what will $x$ be equal to?

*Ans.* $4 + 1$, or 5.

8. If $x - 2$ is equal to 6, what is $x$ equal to?

9. James and John together have 20 oranges; John has 6 less than James: how many has each?

10. A grocer sold 12 pounds of tea and coffee; if the tea be diminished by 3 pounds, and the remainder multiplied by 2, the product is the number of pounds of coffee: how many pounds of each?

11. Ann has a certain number of oranges; Jane has 1 less, and twice her number added to Ann's make 13: how many has each?

ANALYSIS.—Let $x$ denote the number of oranges which Ann has; then, $x - 1$ will denote the number Jane has, and $x + 2x - 2$, or $3x - 2$, will denote the number both have, which is 13. If $3x - 2$ equals 13, $3x$ will be equal to $13 + 2$, or 15; and if $3x$ is equal to 15, $x$ will be equal to 15 divided by 3, which is 5 : hence, Ann has 5 oranges and Jane 4.

WRITTEN.

Let $x$ denote the number Ann has; then,

$$x - 1 = \text{the number Jane has; and}$$
$$2(x - 1) = 2x - 2 = \text{twice what Jane has; also,}$$
$$x + 2x - 2 = 3x - 2 = 13; \text{hence,}$$

$$3x = 13 + 2 = 15; \text{ or } x = \frac{15}{3} = 5.$$

VERIFICATION.

$$5 - 4 = 1; \text{ and } 2 \times 4 + 5 = 13.$$

12. Charles and John have 20 cents, and John has 6 less than Charles: how many has each?

13. James has twice as many oranges as lemons in his basket, and if 5 be taken from the whole number, 19 will remain: how many had he of each?

14. A basket contains apples, peaches, and pears; 29 in all. If 1 be taken from the number of apples, the remainder will denote the number of peaches, and twice that remainder will denote the number of pears: how many are there of each sort?

15. If $2x - 5$ equals 15, what is the value of $x$?

16. If $4x - 5$ is equal to 11, what is the value of $x$?

17. If $5x - 12$ is equal to 18, what is the value of $x$?

18. The sum of two numbers is 32, and the greater exceeds the less by 8 : what are the numbers?

19. The sum of 2 numbers is 9; if the greater number be diminished by 5, and the remainder multiplied by 3, the product will be the less number: what are the numbers?

20. There are three numbers such that 1 taken from the

first will give the second; the second multiplied by 3 will give the third; and their sum is equal to 26: what are the numbers?

21. John and Charles together have just 31 oranges; if 1 be taken from John's, and the remainder be multiplied by 5, the product will be equal to Charles' number: how many has each?

22. A basket is filled with apples, lemons, and oranges, in all 26; the number of lemons exceed the apples by 2, and the number of oranges is double that of the lemons: how many are there of each?

---

## LESSON XI.

1. John has a certain number of apples, the half of which is equal to 10: how many has he?

ANALYSIS.—Let $x$ denote the number of apples; then, $x$ divided by 2 is equal to 10; if one half of $x$ is equal to 10, twice one-half of $x$, or $x$, is equal to twice 10, which is 20; hence, $x$ is equal to 20.

NOTE.—A similar analysis is applicable to any one of the fractional units. Let each question be solved according to the analysis.

2. John has a certain number of oranges, and one-third of his number is 15: how many has he?

3. If one-fifth of a number is 6, what is the number?

4. If one-twelfth of a number is 9, what is the number?

5. What number added to one-half of itself will give a sum equal to 12?

ANALYSIS.—Denote the number by $x$; then, $x$ plus one-half of $x$ equals 12. But $x$ plus one-half of $x$ equals three halves of $x$: hence, three halves of $x$ equal 12. If three halves of $x$ equal 12, one-half of $x$ equals one-third of 12, or 4. If one-half of $x$ equals 4, $x$ equals twice 4, or 8 ' hence, $x$ equals 8.

WRITTEN.

Let $x$ denote the number; then,

$$x + \frac{1}{2}x = \frac{3}{2}x = 12; \text{ then,}$$

$$\frac{1}{2}x = 4, \text{ or } x = 8.$$

VERIFICATION.

$$8 + \frac{8}{2} = 8 + 4 = 12.$$

6. What number added to one-third of itself will give a sum equal to 12?

7. What number added to one-fourth of itself will give a sum equal to 20?

8. What number added to a fifth of itself will make 24?

9. What number diminished by one-half of itself will leave 4? Why?

10. What number diminished by one-third of itself will leave 6?

11. James gave one-seventh of his marbles to William, and then has 24 left: how many had he at first?

12. What number added to two-thirds of itself will give a sum equal to 20?

13. What number diminished by three-fourths of itself will leave 9?

14. What number added to five-sevenths of itself will make 24?

15. What number diminished by seven-eighths of itself will leave 4?

16. What number added to eight-ninths of itself will make 34?

# ELEMENTARY ALGEBRA.

## CHAPTER I,

### DEFINITIONS AND EXPLANATORY SIGNS.

**1.** QUANTITY is anything that can be measured, as number, distance, weight, time, &c.

To measure a thing, is to find how many times it contains some other thing of the same kind, taken as a standard. The assumed standard is called the *unit of measure.*

**2.** MATHEMATICS is the science which treats of the properties and relations of quantities.

In pure mathematics, there are but eight kinds of quantity, and consequently but eight kinds of UNITS, viz.: Units of *Number;* Units of *Currency;* Units of *Length;* Units of *Surface;* Units of *Volume;* Units of *Weight;* Units of *Time;* and Units of *Angular Measure.*

**3.** ALGEBRA is a branch of Mathematics in which the quantities considered are represented by letters, and the operations to be performed are indicated by signs.

---

1. What is quantity? What is the operation of measuring a thing? What is the assumed standard called?

2. What is Mathematics? How many kinds of quantity are there in the pure mathematics? Name the units of those quantities.

3. What is Algebra?

1*

**4.** The quantities employed in Algebra are of two kinds, *Known* and *Unknown*:

> *Known Quantities* are those whose values are given; and
>
> *Unknown Quantities* are those whose values are required.

*Known Quantities* are generally represented by the leading letters of the alphabet, as, *a*, *b*, *c*, &c.

*Unknown Quantities* are generally represented by the final letters of the alphabet; as, *x*, *y*, *z*, &c.

When an unknown quantity becomes known, it is often denoted by the same letter with one or more accents; as, $x'$, $x''$, $x''$. These symbols are read: *x prime; x second; x third*, &c.

**5.** The SIGN OF ADDITION, $+$, is called *plus*. When placed between two quantities, it indicates that the second is to be added to the first. Thus, $a + b$, is read, *a plus b*, and indicates that $b$ is to be added to $a$. If no sign is written, the sign $+$ is understood.

The sign $+$, is sometimes called the *positive* sign, and the quantities before which it is written are called *positive quantities*, or *additive quantities*.

**6.** The SIGN OF SUBTRACTION, $-$, is called *minus*. When placed between two quantities, it indicates that the second is to be subtracted from the first. Thus, the expression,

---

4. How many kinds of quantities are employed in Algebra? How are they distinguished? What are known quantities? What are unknown quantities? By what are the known quantities represented? By what are the unknown quantities represented? When an unknown quantity becomes known, how is it often denoted?

5. What is the sign of addition called? When placed between two quantities, what does it indicate?

6. What is the sign of subtraction called? When placed between two quantities, what does it indicate?

$c - d$, read $c$ minus $d$, indicates that $d$ is to be subtracted from $c$. If $a$ stands for 6, and $d$ for 4, then $a - d$ is equal to 6 — 4, which is equal to 2.

The sign —, is sometimes called the *negative* sign, and the quantities before which it is written are called *negative quantities*, or *subtractive quantities*.

**7.** The SIGN OF MULTIPLICATION, ×, is read, *multiplied by*, or *into*. When placed between two quantities, it indicates that the first is to be multiplied by the second. Thus, $a \times b$ indicates that $a$ is to be multiplied by $b$. If $a$ stands for 7, and $b$ for 5, then, $a \times b$ is equal to 7 × 5, which is equal to 35.

The multiplication of quantities is also indicated by simply writing the letters, one after the other; and sometimes, by placing a point between them; thus,

$a \times b$ signifies the same thing as $ab$, or as $a.b$.

$a \times b \times c$ signifies the same thing as $abc$, or as $a.b.c$.

**8.** A FACTOR is any one of the multipliers of a product. Factors are of two kinds, *numeral* and *literal*. Thus, in the expression, $5abc$, there are four factors : the *numeral* factor, 5, and the three *literal* factors, $a$, $b$, and $c$.

**9.** The SIGN OF DIVISION, ÷, is read, *divided by*. When written between two quantities, it indicates that the first is to be divided by the second.

---

7. How is the sign of multiplication read? When placed between two quantities, what does it indicate? In how many ways may multiplication be indicated?

8. What is a factor? How many kinds of factors are there? How many factors are there in $3abc$?

9. How is the sign of division read? When written between two quantities, what does it indicate? How many ways are there of indicating division?

There are three signs used to denote *division*. Thus,

$a \div b$ denotes that $a$ is to be divided by $b$.

$\dfrac{a}{b}$      denotes that $a$ is to be divided by $b$.

$a \mid b$ denotes that $a$ is to be divided by $b$.

**10.** The Sign of Equality, $=$, is read, *equal to*. When written between two quantities, it indicates that they are equal to each other. Thus, the expression, $a + b = c$, indicates that the sum of $a$ and $b$ is equal to $c$. If $a$ stands for 3, and $b$ for 5, $c$ will be equal to 8.

**11.** The Sign of Inequality, $> <$, is read, *greater than*, or *less than*. When placed between two quantities, it indicates that they are unequal, the greater one being placed at the opening of the sign. Thus, the expression, $a > b$, indicates that $a$ is greater than $b$; and the expression, $c < d$, indicates that $c$ is less than $d$.

**12.** The sign $\therefore$ means, *therefore*, or *consequently*.

**13.** A Coefficient is a number written before a quantity, to show how many times it is taken. Thus,

$$a + a + a + a + a = 5a,$$

in which 5 is the coefficient of $a$.

A coefficient may be denoted either by a *number*, or a *letter*. Thus, $5x$ indicates that $x$ is taken 5 times, and $ax$

---

10. What is the sign of equality? When placed between two quantities, what does it indicate?

11. How is the sign of inequality read? Which quantity is placed on the side of the opening?

12. What does $\therefore$ indicate?

13. What is a coefficient? How many times is $a$ taken in $5a$. By what may a coefficient be denoted? If no coefficient is written, what coefficient is understood? In $5ax$, how many times is $ax$ taken? How many times is $x$ taken?

indicates that $x$ is taken $a$ times. If no coefficient is written, the coefficient 1 is understood. Thus, $a$ is the same as $1a$.

**14.** An Exponent is a number written at the right and above a quantity, to indicate how many times it is taken as a factor. Thus,

$$a \times a \text{ is written } a^2,$$
$$a \times a \times a \quad `` \quad a^3,$$
$$a \times a \times a \times a \quad `` \quad a^4,$$
$$\&c., \qquad \&c.,$$

in which 2, 3, and 4, are *exponents*. The expressions are read, $a$ square, $a$ cube or $a$ third, $a$ fourth; and if we have $a^m$, in which $a$ enters $m$ times as a factor, it is read, $a$ to the $m$th, or simply a, *mth*. The exponent 1 is generally omitted. Thus, $a^1$ is the same as $a$, each denoting that $a$ enters but once as a factor.

**15.** A Power is a product which arises from the multiplication of equal factors. Thus,

$a \times a = a^2$ is the square, or second power of $a$.
$a \times a \times a = a^3$ is the cube, or third power of $a$.
$a \times a \times a \times a = a^4$ is the fourth power of $a$.
$a \times a \times \ldots = a^m$ is the $m$th power of $a$.

**16.** A Root of a quantity is one of the equal factors. The radical sign, $\sqrt{\ }$, when placed over a quantity, indicates that a root of that quantity is to be extracted. The root is indicated by a number written over the radical sign,

14. What is an exponent? In $a^3$, how many times is $a$ taken as a factor? When no exponent is written, what is understood?

15. What is a power of a quantity? What is the third power of 2? Of 4? Of 6?

16. What is the root of a quantity? What indicates a root? What indicates the kind of root? What is the index of the square root? Of the cube root? Of the $m$th root?

called an *index*. When the index is 2, it is generally omitted. Thus,

$\sqrt[2]{a}$, or $\sqrt{a}$, indicates the square root of $a$.

$\sqrt[3]{a}$ indicates the cube root of $a$.

$\sqrt[4]{a}$ indicates the fourth root of $a$.

$\sqrt[m]{a}$ indicates the $m$th root of $a$.

**17.** An ALGEBRAIC EXPRESSION is a quantity written in algebraic language. Thus,

$3a$ $\begin{cases} \text{is the algebraic expression of three times} \\ \text{the number denoted by } a \text{ ;} \end{cases}$

$5a^2$ $\begin{cases} \text{is the algebraic expression of five times} \\ \text{the square of } a \text{ ;} \end{cases}$

$7a^3b^2$ $\begin{cases} \text{is the algebraic expression of seven times} \\ \text{the the cube of } a \text{ multiplied by the} \\ \text{square of } b \text{ ;} \end{cases}$

$3a - 5b$ $\begin{cases} \text{is the algebraic expression of the differ-} \\ \text{ence between three times } a \text{ and five} \\ \text{times } b \text{ ;} \end{cases}$

$2a^2 - 3ab + 4b^2$ $\begin{cases} \text{is the algebraic expression of twice the} \\ \text{square of } a \text{, diminished by three times} \\ \text{the product of } a \text{ by } b \text{, augmented by} \\ \text{four times the square of } b \text{.} \end{cases}$

**18.** A TERM is an algebraic expression of a single quantity. Thus, $3a$, $2ab$, $-5a^2b^2$, are terms.

**19.** The DEGREE of a term is the number of its literal factors. Thus,

$3a$ $\begin{cases} \text{is a term of the first degree, because it contains but} \\ \text{one literal factor.} \end{cases}$

---

17. What is an algebraic expression
18. What is a term?
19. What is the degree of a term? What determines the degree of a term?

$5a^2$ { is of the second degree, because it contains two literal factors.

$7a^3b$ { is of the fourth degree, because it contains four literal factors. The degree of a term is determined by the sum of the exponents of all its letters.

**20.** A MONOMIAL is a single term, unconnected with any other by the signs + or −; thus, $3a^2$, $3b^3a$, are monomials.

**21.** A POLYNOMIAL is a collection of terms connected by the signs + or −; as,

$$3a - 5, \text{ or, } 2a^3 - 3b + 4b^2.$$

**22.** A BINOMIAL is a polynomial of two terms; as,

$$a + b, \ 3a^2 - c^2, \ 6ab - c^2.$$

**23.** A TRINOMIAL is a polynomial of three terms; as,

$$abc - a^3 + c^3, \ ab - gh - f.$$

**24.** HOMOGENEOUS TERMS are those which contain the same number of literal factors. Thus, the terms, $abc$, $- a^3$, $+ c^3$, are homogeneous; as are the terms, $ab$, $- gh$.

**25.** A POLYNOMIAL IS HOMOGENEOUS, when all its terms are homogeneous. Thus, the polynomial, $abc - a^3 + c^3$, is homogeneous; but the polynomial, $ab - gh - f$ is not homogeneous.

**26.** SIMILAR TERMS are those which contain the same literal factors affected with the same exponents. Thus,

$$7ab + 3ab - 2ab,$$

20. What is a monomial?
21. What is a polynomial?
22. What is a binomial?
23. What is a trinomial?
24. What are homogeneous terms?
25. When is a polynomial homogeneous?
26. What are similar terms?

are similar terms; and so also are,

$$4a^2b^2 - 2a^2b^2 - 3a^2b^2;$$

but the terms of the first polynomial and of the last, are not similar.

**27.** THE VINCULUM, ——, the *Bar* |, the *Parenthesis*, ( ), and the *Brackets*, [ ], are each used to connect several quantities, which are to be operated upon in the same manner. Thus, each of the expressions,

$$\overline{a+b+c} \times x, \qquad \begin{array}{c|c} a & x \\ + b & \\ + c & \end{array} \qquad (a+b+c) \times x,$$

and  $[a+b+c] \times x,$

indicates, that the sum of $a$, $b$, and $c$, is to be multiplied by $x$.

**28.** THE RECIPROCAL of a quantity is 1, divided by that quantity; thus,

$$\frac{1}{a}, \quad \frac{1}{a+b}, \quad \frac{c}{d},$$

are the reciprocals of

$$a, \quad a+b, \quad \frac{d}{c}.$$

**29.** THE NUMERICAL VALUE of an algebraic expression, is the result obtained by assigning a numerical value to each letter, and then performing the operations indicated. Thus, the numerical value of the expression,

$$ab + bc + d,$$

when, $a = 1$, $b = 2$, $c = 3$, and $d = 4$, is

$$1 \times 2 + 2 \times 3 + 4 = 12;$$

by performing the indicated operations.

---

27. For what is the vincular used? Point out the other ways in which this may be done?

28. What is the reciprocal of a quantity?

29. What is the numerical value of an algebraical expression?

### EXAMPLES IN WRITING ALGEBRAIC EXPRESSIONS.

1. Write $a$ added to $b$.      *Ans.* $a + b$.
2. Write $b$ subtracted from $a$.      *Ans.* $a - b$.

Write the following:

3. Six times the square of $a$, minus twice the square of $b$.

4. Six times $a$ multiplied by $b$, diminished by 5 times $c$ cube multiplied by $d$.

5. Nine times $a$, multiplied by $c$ plus $d$, diminished by 8 times $b$ multiplied by $d$ cube.

6. Five times $a$ minus $b$, plus 6 times $a$ cube into $b$ cube.

7. Eight times $a$ cube into $d$ fourth, into $c$ fourth, plus 9 times $c$ cube into $d$ fifth, minus 6 times $a$ into $b$, into $c$ square.

8. Fourteen times $a$ plus $b$, multiplied by $a$ minus $b$, plus 5 times $a$, into $c$ plus $d$.

9. Six times $a$, into $c$ plus $d$, minus 5 times $b$, into $a$ plus $c$, minus 4 times $a$ cube $b$ square.

10. Write $a$, multiplied by $c$ plus $d$, plus $f$ minus $g$.

11. Write $a$ divided by $b + c$. Three ways.

12. Write $a - b$ divided by $a + b$.

13. Write a polynomial of three terms; of four terms; of five, of six.

14. Write a homogeneous binomial of the first degree; of the second; of the third; 4th; 5th; 6th.

15. Write a homogeneous trinomial of the first degree; with its second and third terms negative; of the second degree; of the 3rd; of the 4th.

16. Write in the same column, on the slate, or black-board, a monomial, a binomial, a trinomial, a polynomial of four terms, of five terms, of six terms and of seven terms, and all of the *same degree.*

## INTERPRETATION OF ALGEBRAIC LANGUAGE.

Find the numerial values of the following expressions, when,

$$a = 1, \quad b = 2, \quad c = 3, \quad d = 4.$$

1. $ab + bc$.      *Ans.* 8.

2. $a + bc + d$.      *Ans.* 11.

3. $ad + b - c$.      *Ans.* 3.

4. $ab + bc - d$.      *Ans.* 4.

5. $(a + b) c^2 - d$.      *Ans.* 23.

6. $(a + b) (d - b.)$      *Ans.* 6.

7. $(ab + ad) c + d$.      *Ans.* 22.

8. $(ab + c) (ad - a)$.      *Ans.* 15.

9. $3a^2b^2 - 2(a + d + 1)$.      *Ans.* 0.

10. $\dfrac{a + c}{2} \times (a + d)$      *Ans.* 10.

11. $\dfrac{a^2 + b^2 + c^2}{7} \times \dfrac{a^3 + b^3 + c^3 - d}{2}$.      *Ans.* 32.

12. $\dfrac{ab^4 - c - a^3}{6} \times \dfrac{4a^2 - b + d^3}{33}$      *Ans.* 4.

Find the numerical values of the following expressions, when,

$$a = 4, \quad b = 3, \quad c = 2, \quad \text{and } d = 1.$$

13. $\dfrac{a}{2} - \dfrac{b}{3} + c - d$.      *Ans.* 2.

14. $5\left(\dfrac{ab}{3} - \dfrac{a - d}{3}\right)$.      *Ans.* 15.

15. $[(a^2b + 1)d] \div (a^2b + d)$.      *Ans.* 1.

16. $4\left(abc - \dfrac{b^3}{9}\right) \times (30c^3 - ab^3d^3)$.      *Ans.* 11088.

17. $\dfrac{a + b + c}{a - b + d} + \dfrac{abcd}{ab} + \dfrac{4a^2 + b^2 - d^2}{bc + b}$.      *Ans.* $14\frac{1}{5}$.

18. $\dfrac{15(a+d+b)}{3c^2} - \dfrac{a-c}{2} + \dfrac{3}{abd} \times a^3b^3c^3d^3$.      *Ans.* 3465.

# CHAPTER II.

## FUNDAMENTAL OPERATIONS.

## ADDITION.

**30.** Addition is the operation of finding the simplest equivalent expression for the aggregate of two or more algebraic quantities. Such expression is called their Sum.

*When the terms are similar and have like signs.*

**31.** 1. What is the sum of $a$, $2a$, $3a$, and $4a$?
Take the sum of the coefficients, and annex the literal parts. The first term, $a$, has a coefficient, 1, understood (Art. 13).

$$+ \quad a$$
$$+ \quad 2a$$
$$+ \quad 3a$$
$$+ \quad 4a$$
$$\overline{+ \quad 10a}$$

2. What is the sum of $2ab$, $3ab$, $6ab$, and $ab$. When no sign is writtten, the sign $+$ is understood (Art 5).

$$2ab$$
$$3ab$$
$$6ab$$
$$ab$$
$$\overline{12ab}$$

Add the following :

| (3.) | (4.) | (5.) | (6.) |
|------|------|------|------|
| $a$ | $8ab$ | $7ac$ | $+\ 4abc$ |
| $a$ | $7ab$ | $5ac$ | $3abc$ |
| $+\ 2a$ | $15ab$ | $12ac$ | $+\ 7abc$ |

30. What is addition ?
31. What is the rule for addition when the terms are similar and have like signs ?

|  (7.)      |  (8.)    |  (9.)     |  (10.)    |
|------------|----------|-----------|-----------|
| — 3abc     | — 3ad    | — 2adf    | — 9abd    |
| — 2abc     | — 2ad    | — 6adf    | — 15abd   |
| — 5abc     | — 5ad    | — 8adf    | — 24abd   |

Hence, when the terms are similar and have like signs:

### RULE.

*Add the coefficients, and to their sum prefix the common sign ; to this, annex the common literal part.*

### EXAMPLES.

|  (11.)        |  (12.)           |  (13.)                          |
|---------------|------------------|---------------------------------|
| $9ab + ax$    | $8ac^2 - 3b^2$   | $15ab^3c^4 - 12abc^2$           |
| $3ab + 3ax$   | $7ac^2 - 8b^2$   | $12ab^3c^4 - 15abc^2$           |
| $12ab + 4ax$  | $3ac^2 - 9b^2$   | $ab^3c^4 - abc^2$               |

*When the terms are similar and have unlike signs.*

**32.** The signs, $+$ and $-$, stand in direct opposition to each other.

If a merchant writes $+$ before his gains and $-$ before his losses, at the end of the year the sum of the plus numbers will denote the gains, and the sum of the minus numbers the losses. If the gains exceed the losses, the *difference*, which is called the *algebraic sum*, will be plus; but if the losses exceed the gains, the *algebraic sum* will be minus.

1. A merchant in trade gained $1500 in the first quarter of the year, $4000 in the second quarter, but lost $3000 in the third quarter, and $800 in the fourth : what was the result of the year's business?

| 1st quarter, | + 1500   | 3d quarter, | — 3000   |
|--------------|----------|-------------|----------|
| 2d    "      |   3000   | 4th    "    | —  800   |
|              | + 4500   |             | — 3800   |

$$+ 4500 - 3800 = + 700, \text{ or } \$700 \text{ gain.}$$

---

32. What is the rule when the terms are similar and have unlike signs ?

2. A merchant in trade gained $1000 in the first quarter, and $2000 the second quarter; in the third quarter he lost $1500, and in the fourth quarter $1800 : what was the result of the year's business?

| 1st quarter, | $+ 1000$ | 3d quarter | $- 1500$ |
|---|---|---|---|
| 2d    " | $+ 2000$ | 4th    " | $- 1800$ |
|  | $+ 3000$ |  | $- 3300$ |

$$+ 3000 - 3300 = - 300, \text{ or } \$300 \text{ loss.}$$

3. A merchant in the first half-year gained $a$ dollars and lost $b$ dollars; in the second half-year he lost $a$ dollars and gained $b$ dollars : what is the result of the year's business?

| 1st half-year, | $+ a$ | $- b$ |
|---|---|---|
| 2d    " | $- a$ | $+ b$ |
| Result, | $0$ | $0$ |

Hence, *the algebraic sum of a positive and negative quantity is their arithmetical difference, with the sign of the greater prefixed.* Add the following:

| $8ab$ | $4acb^2$ | $- 4a^2b^2c^2$ |
|---|---|---|
| $3ab$ | $- 8acb^2$ | $+ 6a^2b^2c^2$ |
| $- 6ab$ | $acb^2$ | $- 2a^2b^2c^2$ |
| $5ab$ | $- 3acb^2$ | $0$ |

Hence, when the terms are similar and have unlike signs:

I. *Write the similar terms in the same column:*

II. *Add the coefficients of the additive terms, and also the coefficients of the subtractive terms :*

III. *Take the difference of these sums, prefix the sign of the greater, and then annex the literal part.*

1. What is the sum of
$$2a^2b^3 - 5a^2b^3 + 7a^2b^3 + 6a^2b^3 - 11a^2b^3?$$

Having written the similar terms in the same column, we find the sum of the positive coefficients to be 15, and the sum of the negative coefficients to be — 16 : their difference is — 1; hence, the sum is — $a^2b^3$.

$$
\begin{aligned}
&\phantom{-}\ 2a^2b^3 \\
&-\ 5a^2b^3 \\
&+\ 7a^2b^3 \\
&+\ 6a^2b^3 \\
&-\ 11a^2b^3 \\
\hline
&-\ \phantom{1}a^2b^3
\end{aligned}
$$

2. What is the sum of

$3a^2b + 5a^2b - 3a^2b + 4a^2b - 6a^2b - a^2b$?    *Ans.* $2a^2b$.

3. What is the sum of

$12a^3bc^2 - 4a^3bc^2 + 6a^3bc^2 - 8a^3bc^2 + 11a^3bc^2$?    *Ans.* $17a^3bc^2$.

4. What is the sum of

$4a^2b - 8a^2b - 9a^2b + 11a^2b$?    *Ans.* $- 2a^2b$.

5. What is the sum of

$7abc^2 - abc^2 - 7abc^2 + 8abc^2 + 6abc^2$?    *Ans.* $13abc^2$.

6. What is the sum of

$9cb^3 - 5cb^3 - 8ac^2 + 20cb^3 + 9ac^2 - 24cb^3$?    ***Ans.*** $+ ac$.

### To add any Algebraic Quantities.

**33.**    1. What is the sum of $3a$, $5b$, and $- 2c$? Write the quantities, thus,

$$3a + 5b - 2c;$$

which denotes their sum, as there are no *similar terms*.

2. Let it be required to find the sum of the quantities,

$$
\begin{array}{rrr}
2a^2 & 4ab & \\
3a^2 & -\ 3ab & +\ \ b^2 \\
     & 2ab & -\ 5b^2 \\
\hline
5a^2 & -\ 5ab & -\ 4b^2
\end{array}
$$

---

**33.** What is the rule for the addition of any algebraic quantities?

From the preceding examples, we have, for the addition of algebraic quantities, the following

### RULE.

I. *Write the quantities to be added, placing similar terms in the same column, and giving to each its proper sign:*

II. *Add up each column separately and then annex the dissimilar terms with their proper signs.*

### EXAMPLES.

1. Add together the polynomials,

$$3a^2 - 2b^2 - 4ab, \; 5a^2 - b^2 + 2ab, \text{ and } 3ab - 3c^2 - 2b^2.$$

The term $3a^2$ being similar to $5a^2$, we write $8a^2$ for the result of the reduction of these two terms, at the same time slightly crossing them, as in the first term.

$$
\begin{aligned}
3a^2 &- 4ab - 2b^2 \\
5a^2 &+ 2ab - b^2 \\
&+ 3ab - 2b^2 - 3c^2 \\
\hline
8a^2 &+ \phantom{2}ab - 5b^2 - 3c^2
\end{aligned}
$$

Passing then to the term $- 4ab$, which is similar to $+ 2ab$ and $+ 3ab$, the three reduce to $+ ab$, which is placed after $8a^2$, and the terms crossed like the first term. Passing then to the terms involving $b^2$, we find their sum to be $- 5b^2$, after which we write $- 3c^2$.

The marks are drawn across the terms, that none of them may be overlooked and omitted.

| (2.) | (3.) | (4.) |
|---|---|---|
| $7abc + 9ax$ | $8ax + 3b$ | $12a - 6c$ |
| $- 3abc - 3ax$ | $5ax - 9b$ | $- 3a - 9c$ |
| $4abc + 6ax$ | $13ax - 6b$ | $9a - 15c$ |

NOTE.—If $a = 5$, $b = 4$, $c = 2$, $x = 1$, what are the numerical values of the several sums above found?

| (5.) | (6.) | (7.) |
|---|---|---|
| $9a + f$ | $6ax - 8ac$ | $3af + g + m$ |
| $- 6a + g$ | $- 7ax - 9ac$ | $ag - 3af - m$ |
| $- 2a - f$ | $ax + 17ac$ | $ab - ag + 3g$ |

| (8.) | (9.) |
|---|---|
| $7x + 3ab + 3c$ | $8x^2 + 9acx + 13a^2b^2c^2$ |
| $- 3x - 3ab - 5c$ | $- 7x^2 - 13acx + 14a^2b^2c^2$ |
| $5x - 9ab - 9c$ | $- 4x^2 + 4acx - 20a^2b^2c^2$ |

| (10.) | (11.) |
|---|---|
| $22h - 3c - 7f + 3g$ | $19ah^2 + 3a^3b^4 - 8ax^3$ |
| $- 3h + 8c - 2f - 9g + 5x$ | $- 17ah^2 - 9a^3b^4 + 9ax^3$ |

| (12.) | (13.) |
|---|---|
| $7x - 9y + 5z + 3 - g$ | $8a + b$ |
| $- x - 3y - 8 - g$ | $2a - b + c$ |
| $- x + y - 3z + 1 + 7g$ | $- 3a + b + 2d$ |
| $- 2x + 6y + 3z - 1 - g$ | $- 6b - 3c + 3d$ |

14. Add together $- b + 3c - d - 115e + 6f - 5g$, $3b$ $- 2c - 3d - e + 27f$, $5c - 8d + 3f - 7g$, $- 7b - 6c$ $+ 17d + 9e - 5f + 11g$, $- 3b - 5d - 2e + 6f - 9g + h$.
$$Ans. \; - 8b - 109e + 37f - 10g + h.$$

15. Add together the polynomials $7a^2b - 3abc - 8b^2c$ $- 9c^3 + cd^2$, $8abc - 5a^2b + 3c^3 - 4b^2c + cd^2$, and $4a^2b$ $- 8c^3 + 9b^2c - 3d^3$.
$$Ans. \; 6a^2b + 5abc - 3b^2c - 14c^3 + 2cd^2 - 3d^3.$$

16. What is the sum of, $5a^2bc + 6bx - 4af$, $- 3a^2bc$ $- 6bx + 14af$, $- af + 9bx + 2a^2bc$, $+ 6af - 8bx + 6a^2bc$?
$$Ans. \; 10a^2bc + bx + 15af.$$

17. What is the sum of $a^2n^2 + 3a^3m + b$, $- 6a^2n^2$ $- 6a^3m - b$, $+ 9b - 9a^3m - 5a^2n^2$?
$$Ans. \; - 10a^2n^2 - 12a^3m + 9b.$$

18. What is the sum of $4a^3b^2c - 16a^4x - 9ax^3d$, $+ 6a^3b^2c - 6ax^3d + 17a^4x, + 16ax^3d - a^4x - 9a^3b^2c$?
$\qquad\qquad\qquad Ans.\ a^3b^2c + ax^3d.$

19. What is the sum of $- 7g + 3b + 4g - 2b + 3g - 3b + 2b$? $\qquad\qquad Ans.\ 0.$

20. What is the sum of, $ab + 3xy - m - n, - 6xy - 3m + 11n + cd, + 3xy + 4m - 10n + fg$?
$\qquad\qquad\qquad Ans.\ ab + cd + fg.$

21. What is the sum of $4xy + n + 6ax + 9am, - 6xy + 6n - 6ax - 8am, 2xy - 7n + ax - am$? $Ans.\ +ax.$

|  (22.)  |  (23.)  |  (24.)  |
| --- | --- | --- |
| $2(a + b)$ | $5(a^2 - c^2)$ | $9(c^3 - af^3)$ |
| $3(a + b)$ | $- 4(a^2 - c^2)$ | $7(c^3 - af^3)$ |
| $2(a + b)$ | $- 1(a^2 - c^2)$ | $- 10(c^3 - af^3)$ |
| $7(a + b)$ |  | $6(c^3 - af^3)$ |

NOTE. The quantity within the parenthesis must be regarded as a *single* quantity.

25. Add $3a(g^2 - h^2) - 2a(g^2 - h^2) + 4a(g^2 - h^2) + 8a(g^2 - h^2) - 2a(g^2 - h^2)$. $\qquad Ans.\ 11a(g^2 - h^2)$.

26. Add $3c(a^2c - b^2) - 9c(a^2c - b^2) - 7c(a^2c - b^2) + 15c(a^2c - b^2) + c(a^2c - b^2)$. $\qquad Ans.\ 3c(a^2c - b^2)$.

**34.** In algebra, the term *add* does not always, as in arithmetic, convey the idea of augmentation; nor the term *sum*, the idea of a number numerically greater than any of the numbers added. For, if to $a$ we add $- b$, we have, $a - b$, which is, arithmetically speaking, a difference between the number of units expressed by $a$, and the number

34. Do the words *add* and *sum*, in Algebra, convey the same ideas as in Arithmetic. What is the algebraic sum of 9 and $- 4$? Of 8 and $- 2$? May an algebraic sum be negative? What is the sum of 5 and $- 10$? How are such sums distinguished from arithmetical sums?

3

of units expressed by $b$. Consequently, this result is numerically less than $a$. To distinguish this sum from an arithmetical sum, it is called the *algebraic sum*.

---

### SUBTRACTION.

**35.** SUBTRACTION is the operation of finding the difference between two algebraic quantities.

**36.** The quantity to be subtracted is called the *Subtrahend;* and the quantity from which it is taken, is called the *Minuend*.

The *difference* of two quantities, is such a quantity as added to the subtrahend will give a sum equal to the minuend.

### EXAMPLES.

1. From $17a$ take $6a$.

In this example, $17a$ is the minuend, and $6a$ the subtrahend: the difference is $11a$; *because,* $11a$, added to $6a$, gives $17a$.

$$\begin{array}{r} \text{OPERATION.} \\ 17a \\ 6a \\ \hline 11a \end{array}$$

The difference may be expressed by writing the quantities thus:

$$17a - 6a = 11a;$$

in which the sign of the subtrahend is changed from $+$ to $-$.

2. From $15x$ take $-9x$.

The *difference*, or remainder, is such a quantity, as being added to the subtrahend, $-9x$, will give the minuend, $15x$. That quantity is $24x$, and may be found by simply *changing the sign*

$$\begin{array}{r} \text{OPERATION.} \\ 15x \\ -9x \\ \hline 24x \end{array}$$

of the subtrahend, and adding. Whence, we may write,

$$15x - (-9x) = 24x.$$

3. From $10ax$ take $a - b$.

The *difference*, or *remainder*, is such a quantity, as added to $a - b$, will give the minuend, $10ax$: what is that quan tity?

If you change the signs of both terms of the subtrahend, and add, you have, $10ax - a + b$. Is this the true remainder? Certainly. For, if you add the remainder to the subtrahend, $a - b$, you obtain the minuend, $10ax$.

OPERATION.

$$
\begin{array}{r}
10ax \\
+ a - b \\
\hline
\text{Rem. } 10ax - a + b \\
\text{add } + a - b \\
\hline
10ax
\end{array}
$$

It is plain, that if you change the signs of all the terms of the subtrahend, and then add them to the minuend, and to this result add the given subtrahend, the last sum can be no other than the given minuend; hence, the *first* result is the true difference, or remainder (Art. 36).

Hence, for the subtraction of algebraic quantities, we have the following

RULE.

I. *Write the terms of the subtrahend under those of the minuend, placing similar terms in the same column:*

II. *Conceive the signs of all the terms of the subtrahend to be changed from + to −, or from − to +, and then proceed as in Addition.*

EXAMPLES OF MONOMIALS.

|        | (1.) | (2.) | (3.) |
|--------|------|------|------|
| From   | $3ab$ | $6ax$ | $9abc$ |
| take   | $2ab$ | $3ax$ | $7abc$ |
| Rem.   | $ab$ | $3ax$ | $2abc$ |

|              | (4.)            | (5.)            | (6.)            |
|--------------|-----------------|-----------------|-----------------|
| From         | $16a^2b^2c$     | $17a^3b^3c$     | $24a^2b^2x$     |
| take         | $9a^2b^2c$      | $3a^3b^3c$      | $7a^2b^2x$      |
| Rem.         | $7a^2b^2c$      | $14a^3b^3c$     | $17a^2b^2x$     |

|              | (7.)            | (8.)            | (9.)            |
|--------------|-----------------|-----------------|-----------------|
| From         | $3ax$           | $4abx$          | $2am$           |
| take         | $8c$            | $9ac$           | $ax$            |
| Rem.         | $3ax - 8c$      | $4abx - 9ac$    | $2am - ax$      |

10. From $9a^2b^2$ take $3a^2b^2$.      $Ans.\ 6a^2b^2.$

11. From $16a^2xy$ take $-15a^2xy$.      $Ans.\ 31a^2xy.$

12. From $12a^4y^3$ take $8a^4y^3$.      $Ans.\ 4a^4y^3.$

13. From $19a^5x^8y$ take $-18a^5x^8y$.      $Ans.\ 37a^5x^8y.$

14. From $3a^2b^3$ take $3a^3b^2$.      $Ans.\ 3a^2b^3 - 3a^3b^2.$

15. From $7a^2b^4$ take $6a^4b^2$.      $Ans.\ 7a^2b^4 - 6a^4b^2.$

16. From $3ab^2$ take $a^2b^5$.      $Ans.\ 3ab^2 - a^2b^5.$

17. From $x^2y$ take $y^2x$.      $Ans.\ x^2y - y^2x.$

18. From $3x^2y^3$ take $xy$.      $Ans.\ 3x^2y^3 - xy.$

19. From $8a^2y^3x$ take $xyz$.      $Ans.\ 8a^2y^3x - xyz.$

20. From $9a^2b^2$ take $-3a^2b^2$.      $Ans.\ 12a^2b^2.$

21. From $14a^2y^2$ take $-20a^2y^2$.      $Ans.\ 34a^2y^2.$

22. From $-24a^4b^5$ take $16a^4b^5$.      $Ans.\ -40a^4b^5.$

23. From $-13x^2y^4$ take $-14x^2y^4$.      $Ans.\ x^2y^4.$

24. From $-47a^3x^2y$ take $-5a^3x^2y$.    $Ans.\ -42a^3x^2y.$

25. From $-94a^2x^2$ take $3a^2x^2$.      $Ans.\ -97a^2x^2.$

26. From $a + x^2$ take $-y^3$.      $Ans.\ a + x^2 + y^3.$

27. From $a^3 + b^3$ take $-a^3 - b^3$.      $Ans\ 2a^3 + 2b^3.$

28. From $-16a^2x^3y$ take $-19a^2x^3y$.      $Ans.\ +3a^2x^3y.$

29. From $a^2 - x^2$ take $a^2 + x^2$.      $Ans.\ -2x^2.$

GENERAL EXAMPLES.

(1.)

From $6ac - 5ab + c^2$
take $3ac + 3ab + 7c$
Rem. $3ac - 8ab + c^2 - 7c.$

*The same with the signs of the lower line changed.*

(1.)

$6ac - 5ab + c^2$
$-3ac - 3ab - 7c$
$3ac - 8ab + c^2 - 7c.$

(2.)

From $6ax - a + 3b^2$
take $9ax - x + b^2$
Rem. $-3ax - a + x + 2b^2.$

(3.)

$6yx - 3x^2 + 5b$
$yx - 3 + a$
$5yx - 3x^2 + 3 + 5b - a.$

(4.)

From $5a^3 - 4a^2b + 3b^2c$
take $-2a^3 + 3a^2b - 8b^2c$
Rem. $7a^3 - 7a^2b + 11b^2c.$

(5.)

$4ab - cd + 3a^2$
$5ab - 4cd + 3a^2 + 5b^2$
$- ab + 3cd - 5b^2.$

6. From $a + 8$ take $c - 5$.  *Ans.* $a - c + 13$.

7. From $6a^2 - 15$ take $9a^2 + 30$.  *Ans.* $-3a^2 - 45$.

8. From $6xy - 8a^2c^3$ take $-7xy - a^2c^3$.

*Ans.* $13xy - 7a^2c^3$

9. From $a + c$ take $-a - c$.  *Ans.* $2a + 2c$.

10. From $4(a + b)$ take $2(a + b)$.  *Ans.* $2(a + b)$.

11. From $3(a + x)$ take $(a + x)$.  *Ans.* $2(a + x)$.

12. From $9(a^2 - x^2)$ take $-2(a^2 - x^2)$.

*Ans.* $11(a^2 - x^2)$

13. From $6a^2 - 15b^2$ take $-3a^2 + 9b^2$.

*Ans.* $9a^2 - 24b^2$.

14. From $3a^m - 2b^n$ take $a^m - 2b^n$.  *Ans.* $2a^m$.

15. From $9c^2m^2 - 4$ take $4 - 7c^2m^2$. *Ans.* $16c^2m^2 - 8$.

16. From $6am + y$ take $3am - x$. *Ans* $3am + x + y$.

17. From $3ax$ take $3ax - y$.  *Ans.* $+ y$.

18. From $-7f + 3m - 8x$ take $-6f - 5m - 2x + 3d + 8$.    *Ans.* $-f + 8m - 6x - 3d - 8$.

19. From $-a - 5b + 7c + d$ take $4b - c + 2d + 2k$.    *Ans.* $-a - 9b + 8c - d - 2k$.

20. From $-3a + b - 8c + 7e - 5f + 3h - 7x - 13y$ take $k + 2a - 9c + 8e - 7x + 7f - y - 3l - k$.    *Ans.* $-5a + b + c - e - 12f + 3h - 12y + 3l$.

21. From $2x - 4a - 2b + 5$ take $8 - 5b + a + 6x$.    *Ans.* $-4x - 5a + 3b - 3$.

22. From $3a + b + c - d - 10$ take $c + 2a - d$.    *Ans.* $a + b - 10$.

23. From $3a + b + c - d - 10$ take $b - 19 + 3a$.    *Ans.* $c - d + 9$.

24. From $a^3 + 3b^2c + ab^2 - abc$ take $b^3 + ab^2 - abc$.    *Ans.* $a^3 + 3b^2c - b^3$.

25. From $12x + 6a - 4b + 40$ take $4b - 3a + 4x + 6d - 10$.    *Ans.* $8x + 9a - 8b - 6d + 50$.

26. From $2x - 3a + 4b + 6c - 50$ take $9a + x + 6b - 6c - 40$.    *Ans.* $x - 12a - 2b + 12c - 10$.

27. From $6a - 4b - 12c + 12x$ take $2x - 8a + 4b - 6c$.    *Ans.* $14a - 8b - 6c + 10x$.

**38.** In Algebra, the term *difference* does not always, as in Arithmetic, denote a number less than the minuend. For, if from $a$ we subtract $-b$, the remainder will be $a + b$; and this is numerically greater than $a$. We distinguish between the two cases by calling this result the *algebraic difference.*

38. In Algebra, as in Arithmetic, does the term *difference* denote a number less than the minuend? How are the results in the two cases, distinguished from each other?

**39.** When a polynomial is to be subtracted from an algebraic quantity, we inclose it in a parenthesis, place the minus sign before it, and then write it after the minuend. Thus, the expression,

$$6a^2 - (3ab - 2b^2 + 2bc),$$

indicates that the polynomial, $3ab - 2b^2 + 2bc$, is to be taken from $6a^2$. Performing the operations indicated, by the rule for subtraction, we have the equivalent expression :

$$6a^2 - 3ab + 2b^2 - 2bc.$$

The last expression may be changed to the former, by changing the signs of the last three terms, inclosing them in a parenthesis, and prefixing the sign —. Thus,

$$6a^2 - 3ab + 2b^2 - 2bc = 6a^2 - (3ab - 2b^2 + 2bc).$$

In like manner any polynomial may be transformed, as indicated below :

$$7a^3 - 8a^2b - 4b^2c + 6b^3 = 7a^3 - (8a^2b + 4b^2c - 6b^3)$$
$$= 7a^3 - 8a^2b - (4b^2c - 6b^3).$$

$$8a^3 - 7b^2 + c - d = 8a^3 - (7b^2 - c + d)$$
$$= 8a^3 - 7b^2 - (- c + d).$$

$$9b^3 - a + 3a^2 - d = 9b^3 - (a - 3a^2 + d)$$
$$= 9b^3 - a - (- 3a^2 + d).$$

NOTE.—The sign of every quantity is changed when it is placed within a parenthesis, and also when it is brought out.

**40.** From the preceding principles, we have,

$$a - (+ b) = a - b; \text{ and}$$
$$a - (- b) = a + b.$$

39. How is the subtraction of a polynomial indicated? How is this indicated operation performed? How may the result be again put under the first form? What is the general rule in regard to the parenthesis?

40. What is the sign which immediately precedes a quantity called? What is the sign which precedes the parenthesis called? What is the

The sign immediately preceding *b* is called the *sign of the quantity;* the sign preceding the parenthesis is called the *sign of operation;* and the sign resulting from the combination of the signs, is called the *essential sign.*

When the sign of operation is different from the sign of the quantity, the essential sign will be − ; when the sign of operation is the same as the sign of the quantity, the essential sign will be +.

---

## MULTIPLICATION.

**41.** 1. If a man earns *a* dollars in 1 day, how much will he earn in 6 days?

ANALYSIS.—In 6 days he will earn six times as much as in 1 day. If he earns *a* dollars in 1 day, in 6 days he will earn 6*a* dollars.

2. If one hat costs *d* dollars, what will 9 hats cost?
*Ans.* 9*d* dollars.

3. If 1 yard of cloth costs *c* dollars, what will 10 yards cost? *Ans.* 10*c* dollars.

4. If 1 cravat costs *b* cents, what will 40 cost?
*Ans.* 40*b* cents.

5. If 1 pair of gloves costs *b* cents, what will *a* pairs cost?

ANALYSIS.—If 1 pair of gloves cost *b* cents, *a* pairs will cost as many times *b* cents as there are units in *a*: that is, *b* taken *a* times, or *ab*; which denotes the *product* of *b* by *a*, or of *a* by *b*.

---

resulting sign called? When the sign of operation is different from the sign of the quantity, what is the essential sign? When the sign of operation is the same as the sign of the quantity, what is the essential sign?

41. What is Multiplication? What is the quantity to be multiplied called? What is that called by which it is multiplied? What is the result called?

MULTIPLICATION *is the operation of finding the product of two quantities.*

The quantity to be multiplied is called the *Multiplicand;* that by which it is multiplied is called the *Multiplier;* and the result is called the *Product.* The Multiplier and Multiplicand are called *Factors* of the Product.

6. If a man's income is $3a$ dollars a week, how much will he receive in $4b$ weeks?

$$3a \times 4b = 12ab.$$

If we suppose $a = 4$ dollars, and $b = 3$ weeks, the product will be 144 dollars.

NOTE.—It is proved in Arithmetic (Davies' School, Art. 48. University, Art. 50), that the product is not altered by changing the arrangement of the factors; that is,

$$12ab = a \times b \times 12 = b \times a \times 12 = a \times 12 \times b.$$

### MULTIPLICATION OF POSITIVE MONOMIALS.

**42.** Multiply $3a^2b^2$ by $2a^2b$. We write,

$$3a^2b^2 \times 2a^2b = 3 \times 2 \times a^2 \times a^2 \times b^2 \times b$$
$$= 3 \times 2\,a\,a\,a\,a\,b\,b\,b;$$

in which $a$ is a factor 4 times, and $b$ a factor 3 times; hence (Art. 14),

$$3a^2b^2 \times 2a^2b = 3 \times 2a^4b^3 = 6a^4b^3,$$

in which *we multiply the coefficients together, and add the exponents of the like letters.*

The product of any two positive monomials may be found in like manner; hence the

#### RULE.

I. *Multiply the coefficients together for a new coefficient:*

II. *Write after this coefficient all the letters in both mono-*

---

42. What is the rule for multiplying one monomial by another?

3*

*mials, giving to each letter an exponent equal to the sum of its exponents in the two factors.*

### EXAMPLES.

1.        $8a^2bc^2 \times 7abd^2 = 56a^3b^2c^2d^2.$

2.        $21a^3b^2cd \times 8abc^3 = 168a^4b^3c^4d.$

3        $4abc \times 7df = 28abcdf.$

| | (4.) | (5.) | (6.) |
|---|---|---|---|
| Multiply | $3a^2b$ | $12a^2x$ | $6xyz$ |
| by | $2a^2b$ | $12x^2y$ | $ay^2z$ |
| | $6a^4b^2$ | $144a^2x^3y$ | $6axy^3z^2$ |

| (7.) | (8.) | (9.) |
|---|---|---|
| $a^2xy$ | $3ab^2c^3$ | $87ax^2y$ |
| $2xy^2$ | $9a^2b^3c$ | $3b^3x^4y^3$ |
| $2a^2x^2y^3$ | $27a^3b^5c^4$ | $261ab^3x^6y^4$ |

10. Multiply $5a^3b^2x^2$ by $6c^5x^6.$        *Ans.* $30a^3b^2c^5x^8.$

11. Multiply $10a^4b^5c^8$ by $7acd.$        *Ans.* $70a^5b^5c^9d.$

12. Multiply $36a^8b^7c^6d^5$ by $20ab^2c^3d^4.$ *Ans.* $720a^9b^9c^9d^9.$

13. Multiply $5a^m$ by $3ab^n.$        *Ans.* $15a^{m+1}b^n.$

14. Multiply $3a^mb^3$ by $6a^2b^n.$        *Ans.* $18a^{m+2}b^{n+3}.$

15. Multiply $6a^mb^n$ by $9a^5b^7.$        *Ans.* $54a^{m+5}b^{n+7}.$

16. Multiply $5a^mb^n$ by $2a^pb^q.$        *Ans.* $10a^{m+p}b^{n+q}.$

17. Multiply $5a^mb^2c^2$ by $2ab^nc.$        *Ans.* $10a^{m+1}b^{n+2}c^3.$

18. Multiply $6a^2b^mc^n$ by $3a^3b^2c^2.$        *Ans.* $18a^5b^{m+2}c^{n+2}.$

19. Multiply $20a^5b^5cd$ by $12a^2x^2y.$ *Ans.* $240a^7b^5cdx^2y.$

20. Multiply $14a^4b^6d^4y$ by $20a^3c^2x^2y.$ *A.* $280a^7b^6c^2d^4x^2y^2.$

21. Multiply $8a^3b^3y^4$ by $7a^4bxy^5.$        *Ans.* $56a^7b^4xy^9.$

22. Multiply $75axyz$ by $5a^5bcdx^2y^2.$ *Ans.* $375a^6bcdx^3y^3z.$

23. Multiply $64a^3m^5x^4yz$ by $8ab^2c^3$. $A$. $512a^4b^2c^3m^5x^4yz$.
24. Multiply $9a^2b^2c^2d^3$ by $12a^3b^4c^6$. *Ans.* $108a^5b^6c^8d^3$.
25. Multiply $216ab^7c^3d^8$ by $3a^3b^2c^5$. *Ans.* $648a^4b^9c^8d^8$.
26. Multiply $70a^8b^7c^4d^2fx$ by $12a^7b^5c^3dx^2y^3$.

$$\textit{Ans.}\ \ 840a^{15}b^{12}c^7d^3fx^3y^3.$$

### MULTIPLICATION OF POLYNOMIALS.

**43.** 1. Multiply $a - b$ by $c$.

It is required to take the *difference* between $a$ and $b$, $c$ times; or, to take $c$, $a - b$ times.

$$\begin{array}{r} a - b \\ c \\ \hline ac - bc \end{array}$$

As we can not subtract $b$ from $c$, we begin by taking $a$, $c$ times, which is $ac$; but this product is too large by $b$ taken $c$ times, which is $bc$; hence, the true product is $ac - bc$.

$$\begin{array}{r} 8 - 3 = 5 \\ 7 \ \ldots \ 7 \\ \hline 56 - 21 = 35 \end{array}$$

If $a$, $b$, and $c$, denote numbers, as $a = 8$, $b = 3$, and $c = 7$, the operation may be written in figures.

Multiply $a - b$ by $c - d$.

It is required to take $a - b$ as many times as there are units in $c - d$.

$$\begin{array}{r} a - b \\ c - d \\ \hline ac - bc \\ - ad + bd \\ \hline ac - bc - ad + bd \end{array}$$

If we take $a - b$, $c$ times, we have $ac - bc$; but this product is too large by $a - b$ taken $d$ times. But $a - b$ taken $d$ times, is $ad - db$. Subtracting this product from the preceding, by changing the signs of its terms (Art. 37), and we have,

$$\begin{array}{r} 8 - 3 \ \ \ = 5 \\ 7 - 2 \ \ \ = 5 \\ \hline 56 - 21 \\ - 16 + 6 \\ \hline 56 - 37 + 6 = 25. \end{array}$$

$$(a - b) \times (a - c) = ab - bc - ad + bd.$$

Hence, we have the following

I. *When the factors have like signs, the sign of their product will be $+$ :*

II. *When the factors have unlike signs, the sign of their product will be $-$ :*

Therefore, we say in Algebraic language, that $+$ multiplied by $+$, or $-$ multiplied by $-$, gives $+$; $-$ multiplied by $+$, or $+$ multiplied by $-$, gives $-$.

Hence, for the multiplication of polynomials, we have the following

RULE.

*Multiply every term of the multiplicand by each term of the multiplier, observing that like signs give $+$, and unlike signs $-$ ; then reduce the result to its simplest form.*

EXAMPLES IN WHICH ALL THE TERMS ARE PLUS.

1. 'Multiply . . . . . $3a^2 + 4ab + b^2$
    by . . . . . $2a + 5b$

$$6a^3 + 8a^2b + 2ab^2$$

The product, after reducing,     $+ 15a^2b + 20ab^2 + 5b^3$

becomes . . . . $6a^3 + 23a^2b + 22ab^2 + 5b^3$.

**44.** NOTE.—It will be found convenient to *arrange* the terms of the polynomials with reference to some letter; that is, to write them down, so that the highest power of that letter shall enter the first term; the next highest, the second term, and so on to the last term.

---

44. How are the terms of a polynomial arranged with reference to a particular letter? What is this letter called? I 'the leading letter in the multiplicand and multiplier is the same, which will be the leading letter in the product?

The letter with reference to which the arrangement is made, is called the *leading letter*. In the above example the leading letter is $a$. The leading letter of the product will always be the same as that of the factors.

2. Multiply $x^2 + 2ax + a^2$ by $x + a$.
   > *Ans.* $x^3 + 3ax^2 + 3a^2x + a^3$.

3. Multiply $x^3 + y^3$ by $x + y$.
   > *Ans.* $x^4 + xy^3 + x^3y + y^4$.

4. Multiply $3ab^2 + 6a^2c^2$ by $3ab^2 + 3a^2c^2$.
   > *Ans.* $9a^2b^4 + 27a^3b^2c^2 + 18a^4c^4$.

5. Multiply $a^2b^2 + c^2d$ by $a + b$.
   > *Ans.* $a^3b^2 + ac^2d + a^2b^3 + bc^2d$.

6. Multiply $3ax^2 + 9ab^3 + cd^5$ by $6a^2c^2$.
   > *Ans.* $18a^3c^2x^2 + 54a^3c^2b^3 + 6a^2c^3d^5$.

7. Multiply $64a^3x^3 + 27a^2x + 9ab$ by $8a^3cd$.
   > *Ans.* $512a^6cdx^3 + 216a^5cdx + 72a^4bcd$.

8. Multiply $a^3 + 3a^2x + 3ax^2 + x^3$ by $a + x$.
   > *Ans.* $a^4 + 4a^3x + 6a^2x^2 + 4ax^3 + x^4$.

9. Multiply $x^2 + y^2$ by $x + y$.
   > *Ans.* $x^3 + xy^2 + x^2y + y^3$.

10. Multiply $x^5 + xy^6 + 7ax$ by $ax + 5ax$.
    > *Ans.* $6ax^6 + 6ax^2y^6 + 42a^2x^2$.

11. Multiply $a^3 + 3a^2b + 3ab^2 + b^3$ by $a + b$.
    > *Ans.* $a^4 + 4a^3b + 6a^2b^2 + 4ab^3 + b^4$.

12. Multiply $x^3 + x^2y + xy^2 + y^3$ by $x + y$.
    > *Ans.* $x^4 + 2x^3y + 2x^2y^2 + 2xy^3 + y^4$.

13. Multiply $x^3 + 2x^2 + x + 3$ by $3x + 1$.
    > *Ans.* $3x^4 + 7x^3 + 5x^2 + 10x + 3$.

### GENERAL EXAMPLES.

1. Multiply . . . . . . . $2ax - 3ab$
   by . . . . . . . . $3x - b.$

The product . . . . . . . $6ax^2 - 9abx$

becomes after . . . . . . , $\qquad - 2abx + 3ab^2$

reducing . . . . . . . . $6ax^2 - 11abx + 3ab^2.$

2. Multiply $a^4 - 2b^3$ by $a - b.$
   $\qquad$ Ans. $a^5 - 2ab^3 - a^4b + 2b^4.$

3. Multiply $x^2 - 3x - 7$ by $x - 2.$
   $\qquad$ Ans. $x^3 - 5x^2 - x + 14.$

4. Multiply $3a^2 - 5ab + 2b^2$ by $a^2 - 7ab.$
   $\qquad$ Ans. $3a^4 - 26a^3b + 37a^2b^2 - 14ab^3.$

5. Multiply $b^2 + b^4 + b^6$ by $b^2 - 1.$ $\quad$ Ans. $b^8 - b^2.$

6. Multiply $x^4 - 2x^3y + 4x^2y^2 - 8xy^3 + 16y^4$ by $x + 2y.$
   $\qquad$ Ans. $x^5 + 32y^5.$

7. Multiply $4x^2 - 2y$ by $2y.$ $\qquad$ Ans. $8x^2y - 4y^2.$

8. Multiply $2x + 4y$ by $2x - 4y.$ $\quad$ Ans. $4x^2 - 16y^2.$

9. Multiply $x^3 + x^2y + xy^2 + y^3$ by $x - y.$
   $\qquad$ Ans. $x^4 - y^4.$

10. Multiply $x^2 + xy + y^2$ by $x^2 - xy + y^2.$
    $\qquad$ Ans. $x^4 + x^2y^2 + y^4.$

11. Multiply $2a^2 - 3ax + 4x^2$ by $5a^2 - 6ax - 2x^2.$
    $\qquad$ Ans. $10a^4 - 27a^3x + 34a^2x^2 - 18ax^3 - 8x^4.$

12. Multiply $3x^2 - 2xy + 5$ by $x^2 + 2xy - 3.$
    $\qquad$ Ans. $3x^4 + 4x^3y - 4x^2 - 4x^2y^2 + 16xy - 15.$

13. Multiply $3x^3 + 2x^2y^2 + 3y^2$ by $2x^3 - 3x^2y^2 + 5y^3.$
    $\qquad$ Ans. $\begin{cases} 6x^6 - 5x^5y^2 - 6x^4y^4 + 6x^3y^2 + \\ 15x^3y^3 - 9x^2y^4 + 10x^2y^5 + 15y^5. \end{cases}$

14. Multiply $8ax - 6ab - c$ by $2ax + ab + c.$
    Ans. $16a^2x^2 - 4a^2bx - 6a^2b^2 + 6acx - 7abc - c^2.$

15. Mult.ply $3a^2 - 5b^2 + 3c^2$ by $a^2 - b^2$.

*Ans.* $3a^4 - 8a^2b^2 + 3a^2c^2 + 5b^4 - 3b^2c^2$.

16. $\qquad 3a^2 - 5bd + cf$
$\qquad - 5a^2 + 4bd - 8cf$.

Pro.red. $- 15a^4 + 37a^2bd - 29a^2cf - 20b^2d^2 + 44bcdf - 8c^2f^2$

17. Multiply $a^mx - a^2b^2$ by $a^2x^n$.

*Ans.* $a^{m+2}x^{n+1} - a^4b^2x^n$.

18. Multiply $a^m + b^n$ by $a^m - b^n$.   *Ans.* $a^{2m} - b^{2n}$.

19. Multiply $a^m + b^n$ by $a^m + b^n$.

*Ans.* $a^{2m} + 2a^mb^n + b^{2n}$.

---

## DIVISION.

**45.** Division is the operation of finding from two quantities a third, which being multiplied by the second, will produce the first.

The first is called the *Dividend*, the second the *Divisor*, and the third, the *Quotient*.

Division is the converse of Multiplication. In it, we have given the product and one factor, to find the other. The rules for Division are just the converse of those for Multiplication.

*To divide one monomial by another.*

**46.** Divide $72a^5$ by $8a^3$. The division is indicated, thus:

$$\frac{72a^5}{8a^3}.$$

The quotient must be such a monomial, as, being *multiplied by the divisor*, will give the dividend. Hence, the coefficient

---

45. What is division? What is the first quantity called? The second? The third? What is given in division? What is required?

46. What is the rule for the division of monomials?

of the quotient must be 9, and the literal part $a^2$; for these quantities multiplied by $8a^3$ will give $72a^5$. Hence,

$$\frac{72a^5}{8a^3} = 9a^2.$$

The coefficient 9 is obtained by dividing 72 by 8; and the literal part is found by giving to $a$, an exponent equal to 5 minus 3.

Hence, for dividing one monomial by another, we have the following

RULE.

I. *Divide the coefficient of the dividend by the coefficient of the divisor, for a new coefficient:*

II. *After this coefficient write all the letters of the dividend, giving to each an exponent equal to the excess of its exponent in the dividend over that in the divisor.*

SIGNS IN DIVISION.

**47.** Since the Quotient multiplied by the Divisor must produce the Dividend: and, since the product of two factors having the *same sign* will be + ; and the product of two factors having different signs will be — ; we conclude:

1. When the signs of the dividend and divisor are like, the sign of the quotient will be +.

2. When the signs of the dividend and divisor are unlike, the sign of the quotient will be —. Again, for brevity, we say,

+ divided by +, and — divided by —, give + ;
— divided by +, and + divided by —, give —.

$$\frac{+\,ab}{+\,a} = +\,b; \qquad \frac{-\,ab}{-\,b} = +\,a;$$

$$\frac{-\,ab}{+\,a} = -\,b; \qquad \frac{+\,ab}{-\,a} = -\,b.$$

47. What is the rule for the signs, in division?

EXAMPLES.

(1.)

$$+ \frac{18a^5b^2c}{9a^3bc} = + 2a^2b.$$

(2.)

$$\frac{- 15a^3x^2y}{- 5a^2x} = + 3axy.$$

(3.)

$$\frac{- 24a^4bc}{+ 3abc} = - 8a^3.$$

(4.)

$$\frac{32a^8b^3x^5}{- 8a^7b^3x} = - 4ax^4.$$

5. Divide $15ax^2y^3$ by $- 3ay$.   $Ans.\ - 5x^2y^2.$

6. Divide $84ab^3x$ by $12b^2$.   $Ans.\ 7abx.$

7. Divide $- 36a^4b^5c^2$ by $9a^3b^2c$.   $Ans.\ - 4ab^3c.$

8. Divide $- 99a^4b^4x^5$ by $11a^3b^2x^4$.   $Ans.\ - 9ab^2x.$

9. Divide $108x^6y^5z^3$ by $54x^5z$.   $Ans.\ 2xy^5z^2.$

10. Divide $64x^7y^5z^6$ by $- 16x^6y^4z^5$.   $Ans.\ - 4xyz.$

11. Divide $- 96a^7b^6c^5$ by $12a^2bc$.   $Ans.\ - 8a^5b^5c^4.$

12. Divide $- 38a^4b^6d^4$ by $2a^3b^5d$.   $Ans.\ - 19abd^3.$

13. Divide $- 64a^5b^4c^8$ by $32a^4bc$.   $Ans.\ - 2ab^3c^7.$

14. Divide $128a^5x^6y^7$ by $16axy^4$.   $Ans.\ 8a^4x^5y^3.$

15. Divide $- 256a^4b^9c^8d^7$ by $16a^3bc^6$. $Ans.\ - 16ab^8c^2d^7.$

16. Divide $200a^8m^2n^2$ by $- 50a^7mn$.   $Ans.\ - 4amn.$

17. Divide $300x^3y^4z^2$ by $60xy^2z$.   $Ans.\ 5x^2y^2z.$

18. Divide $27a^5b^2c^2$ by $- 9abc$.   $Ans.\ - 3a^4bc.$

19. Divide $64a^3y^6z^8$ by $32ay^5z^7$.   $Ans.\ 2a^2yz.$

20. Divide $- 88a^5b^6c^8$ by $11a^3b^4c^6$.   $Ans.\ - 8a^2b^2c^2.$

21. Divide $77a^4y^3z^4$ by $- 11a^4y^3z^4$.   $Ans.\ - 7.$

22. Divide $84a^4b^2c^2d$ by $- 42a^4b^2c^2d$.   $Ans.\ - 2.$

23. Divide $- 88a^6b^7c^6$ by $8a^5b^6c^6$.   $Ans.\ - 11ab.$

24. Divide $16x^2$ by $- 8x$.   $Ans.\ - 2x.$

25. Divide $- 88a^nb^2$ by $11a^mb$.   $Ans.\ - 8a^{n-m}b.$

26. Divide $77a^m b^n$ by $-11a^2 b^3$.     Ans. $-7a^{m-2}b^{n-3}$.

27. Divide $84a^8 b^m$ by $42a^n b^9$.     Ans. $2a^{8-n}b^{m-9}$.

28. Divide $-88a^p b^q$ by $8a^n b^m$.     Ans. $-11a^{p-n}b^{q-m}$.

29. Divide $96ab^p$ by $48a^n b^q$.     Ans. $2a^{1-n}b^{p-q}$.

30. Divide $168x^a y^b$ by $12x^n y^m$.     Ans. $14x^{a-n}y^{b-m}$.

31. Divide $256ab^3 c^2$ by $16a^n b^m c^p$.     Ans. $16a^{1-n}b^{3-m}c^{2-p}$.

## MONOMIAL FRACTIONS.

**48.** It follows from the preceding rules, that the exact division of monomials will be impossible:

1st. When the coefficient of the dividend is not exactly divisible by that of the divisor.

2d. When the exponent of the same letter is greater in the divisor than in the dividend.

3d. When the divisor contains one or more letters not found in the dividend.

In either case, the quotient will be expressed by a fraction.

A fraction is said to be in its *simplest form*, when the numerator and denominator do not contain a common factor.

For example, $12a^4 b^2 cd$, divided by $8a^2 bc^2$, gives

$$\frac{12a^4 b^2 d}{8a^2 bc^2};$$

which may be reduced by dividing the numerator and denominator by the common factors, 4, $a^2$, $b$, and $c$, giving

$$\frac{12a^4 b^2 cd}{8a^2 bc^2} = \frac{3a^2 bd}{2c}.$$

Also,     $$\frac{25a^5 b^2 d^3}{15a^4 b^6 d^4} = \frac{5a}{3b^4 d}.$$

---

48. Under what circumstances will the division of monomials be impossible? How will the quantities then be expressed? How is a monomial fraction reduced to *its* simplest form?

Hence, for the reduction of a monomial fraction to its simplest form, we have the following

RULE.

*Suppress every factor, whether numerical or literal, that is common to both terms of the fraction ; the result will be the reduced fraction sought.*

EXAMPLES.

( 1.)    ( 2.)

$$\frac{48a^3b^2cd^3}{36a^2b^3c^2de} = \frac{4ad^2}{3bce} : \text{ and } \frac{37ab^3c^5d}{6a^3bc^4d^2} = \frac{37b^2c}{6a^2d} ;$$

( 3.)    ( 4.)

$$\text{also,} \quad \frac{7a^2b}{14a^3b^2} = \frac{1}{2ab} : \text{ and } \frac{4a^2b^2}{6ab^4} = \frac{2a}{3b^2} .$$

5. Divide $49a^2b^2c^6$ by $14a^3bc^4$.    *Ans.* $\dfrac{7bc^2}{2a}$.

6. Divide $6amn$ by $3abc$.    *Ans.* $\dfrac{2mn}{bc}$.

7. Divide $18a^2b^2mn^2$ by $12a^4b^4cd$.    *Ans.* $\dfrac{3mn^2}{2a^2b^2cd}$.

8. Divide $28a^5b^6c^7d^8$ by $16ab^9cd^7m$.    *Ans.* $\dfrac{7a^4c^6d}{4b^3m}$.

9. Divide $72a^3c^2b^2$ by $12a^5c^4b^3d$.    *Ans.* $\dfrac{6}{a^2c^2bd}$.

10. Divide $100a^8b^5xmn$ by $25a^3b^4d$.    *Ans.* $\dfrac{4a^5bxmn}{d}$.

11. Divide $96a^5b^8c^9df$ by $75a^2cxy$.    *Ans.* $\dfrac{32a^3b^8c^8df}{25xy}$.

12. Divide $85m^2n^3fx^2y^3$ by $15am^4nf$.    *Ans.* $\dfrac{17n^2x^2y^3}{3am^2}$.

13. Divide $127d^3x^2y^2$ by $16d^4x^4y^4$.    *Ans.* $\dfrac{127}{16dx^2y^2}$.

68 ELEMENTARY ALGEBRA.

**49.** In dividing monomials, it often happens that the exponents of the same letter, in the dividend and divisor, are equal; in which case that letter may not appear in the quotient. It might, however, be retained by giving to it the exponent 0.

If we have expressions of the form

$$\frac{a}{a},\ \frac{a^2}{a^2},\ \frac{a^3}{a^3},\ \frac{a^4}{a^4},\ \frac{a^5}{a^5},\ \&c.,$$

and apply the rule for the exponents, we shall have,

$$\frac{a}{a}=a^{1-1}=a^0,\ \frac{a^2}{a^2}=a^{2-2}=a^0,\ \frac{a^3}{a^3}=a^{3-3}=a^0,\ \&c.$$

But since any quantity divided by itself is equal to 1, it follows that,

$$\frac{a}{a}=a^0=1,\ \frac{a^2}{a^2}=a^{2-2}=a^0=1,\ \&c.;$$

or, finally, if we designate the exponent by $m$, we have,

$$\frac{a^m}{a^m}=a^{m-m}=a^0=1;\ \text{that is,}$$

*The 0 power of any quantity is equal to 1:* therefore,
*Any quantity may be retained in a term, or introduced into a term, by giving it the exponent 0.*

### EXAMPLES.

1. Divide $6a^2b^2c^4$ by $2a^2b^2$.

$$\frac{6a^2b^2c^4}{2a^2b^2}=3a^{2-2}b^{2-2}c^4=3a^0b^0c^4=3c^4.$$

2. Divide $8a^4b^3c^5$ by $-4a^4b^3c$. *Ans.* $-2a^0b^0c^4=-2c^4.$

3. Divide $-32m^3n^2x^2y^2$ by $4m^3n^2xy$.

*Ans.* $-8m^0n^0xy=-8xy.$

**4.** Divide $-96a^1b^5c^n$ by $-24a^4b^5$. *Ans.* $4a^0b^0c^n = 4c^n$.

**5.** Introduce $a$, as a factor, into $6b^5c^4$. *Ans.* $6a^0b^5c^4$.

**6.** Introduce $ab$, as factors, into $9c^5d^n$. *Ans.* $9a^0b^0c^5d^n$.

**7.** Introduce $abc$, as factors, into $8d^1f^m$. *A.* $8a^0b^0c^0d^1f^m$.

**50.** When the exponent of any letter is greater in the divisor than it is in the dividend, the exponent of that letter in the quotient may be written with a negative sign. Thus,

$$\frac{a^2}{a^5} = \frac{1}{a^3}; \text{ also, } \frac{a^2}{a^5} = a^{2-5} = a^{-3}, \text{ by the rule;}$$

hence, $$a^{-3} = \frac{1}{a^3}.$$

Since, $a^{-3} = \frac{1}{a^3}$, we have, $b \times a^{-3} = \frac{b}{a^3}$;

that is, $a$ in the numerator, with a negative exponent, is equal to $a$ in the denominator, with an equal positive exponent; hence,

*Any quantity having a negative exponent, is equal to the reciprocal of the same quantity with an equal positive exponent.*

Hence, also,

*Any factor may be transferred from the denominator to the numerator of a fraction, or the reverse, by changing the sign of its exponent.*

EXAMPLES.

1. Divide $32a^2bc$ by $16a^5b^2$.

$$\textit{Ans. } \frac{32a^2bc}{16a^5b^2} = 2a^{-3}b^{-1}c = \frac{2c}{a^3b}.$$

---

50. When the exponent of any letter in the divisor is greater than in the dividend, how may the exponent of that letter be written in the quotient? What is a quantity with a negative exponent equal to? How may a factor be transferred from the numerator to the denominator of a fraction?

2. $\dfrac{54a^2b^3c}{9a^4b^5} = 6a^{-2}b^{-2}c$.        *Ans.* $\dfrac{6c}{a^2b^2}$.

3. Reduce $\dfrac{17x^2y^3z}{51x^4y^3}$.        *Ans.* $\dfrac{x^{-2}z}{3}$, or $\dfrac{z}{3x^2}$

4. In $5ay^{-3}x^{-2}$, get rid of the negative exponents.

             *Ans.* $\dfrac{5a}{y^3x^2}$.

5. In $\dfrac{4a^2b^3x^{-2}}{3a^{-3}b^{-5}}$, get rid of the negative exponents.

             *Ans.* $\dfrac{4a^5b^8}{3x^2}$.

6. In $\dfrac{15a^{-3}c^{-4}d^{-5}}{45x^{-3}y^{-5}c^{-2}}$, get rid of the negative exponents.

             *Ans.* $\dfrac{x^3y^5c^2}{3a^3c^4d^5}$.

7. Reduce $\dfrac{-8a^{-3}b^5c}{14a^2b^{-5}c}$.    *Ans.* $\dfrac{-4a^{-5}b^{10}c^0}{7}$, or $\dfrac{-4b^{10}}{7a^5}$.

8. Reduce $72a^5b^2 \div 8a^6b^3$.     *Ans.* $9a^{-1}b^{-1}$, or $\dfrac{9}{ab}$.

9. In $\dfrac{15a^{-4}b^6c^{-1}}{5a^{-2}b^{-1}}$, get rid of the negative exponents.

             *Ans.* $\dfrac{3b^7}{a^2c}$.

10. Reduce $\dfrac{-15a^{-5}b^{-5}c^2}{-5a^{-6}b^{-7}}$.     *Ans.* $3ab^2c^2$.

## To divide a polynomial by a monomial.

*' 51.* To divide a polynomial by a monomial:

*Divide each term of the dividend, separately, by the divisor ; the algebraic sum of the quotients will be the quotient sought.*

### EXAMPLES.

1. Divide $3a^2b^2 - a$ by $a$.       *Ans.* $3ab^2 - 1$.

---

51. How do you divide a polynomial by a monomial?

2. Divide $5a^3b^2 - 25a^4b^2$ by $5a^3b^2$.    *Ans.* $1 - 5a$.

3. Divide $35a^2b^2 - 25ab$ by $- 5ab$.    *Ans.* $-\cdot 7ab + 5$.

4. Divide $10ab - 15ac$ by $5a$.    *Ans.* $2b - 3c$.

5. Divide $6ab - 8ax + 4a^2y$ by $2a$.

                             *Ans.* $3b - 4x + 2ay$.

6. Divide $- 15ax^2 + 6x^3$ by $- 3x$.    *Ans.* $5ax - 2x^2$.

7. Divide $- 21xy^2 + 35a^2b^2y - 7c^2y$ by $- 7y$.

                             *Ans.* $3xy - 5a^2b^2 + c^2$.

8. Divide $40a^8b^4 + 8a^4b^7 - 32a^4b^4c^4$ by $8a^4b^4$.

                             *Ans.* $5a^4 + b^3 - 4c^4$.

## DIVISION OF POLYNOMIALS.

**52.** 1. Divide $- 2a + 6a^2 - 8$ by $2 + 2a$.

$$\begin{array}{ll} Dividend. & Divisor. \\ 6a^2 - 2a - 8\ |\ 2a + 2 & \\ \underline{6a^2 + 6a}\ \ \ \ \ \overline{3a - 4} & Quotient. \\ \ \ \ \ \ -8a - 8 & \\ \ \ \ \ \ \underline{-8a - 8} & \\ \ \ \ \ \ \ \ \ \ 0 \ \ \ \ \ \ Remainder. & \end{array}$$

We first *arrange* the dividend and divisor with reference to $a$ (Art. 44), placing the divisor on the left of the dividend. Divide the first term of the dividend by the first term of the divisor; the result will be the first term of the quotient, which, for convenience, we place under the divisor. The product of the divisor by this term ($6a^2 + 6a$), being subtracted from the dividend, leaves a new dividend, which may be treated in the same way as the original one, and so on to the end of the operation.

---

52. What is the rule for dividing one polynomial by another? When is the division exact? When is it not exact?

Since all similar cases may be treated in the same way, we have, for the division of polynomials, the following

RULE.

I. *Arrange the dividend and divisor with reference to the same letter:*

II. *Divide the first term of the dividend by the first term of the divisor, for the first term of the quotient. Multiply the divisor by this term of the quotient, and subtract the product from the dividend:*

III. *Divide the first term of the remainder by the first term of the divisor, for the second term of the quotient. Multiply the divisor by this term, and subtract the product from the first remainder, and so on:*

IV. *Continue the operation, until a remainder is found equal to 0, or one whose first term is not divisible by that of the divisor.*

Note.—1. When a remainder is found equal to 0, the division is exact.

2. When a remainder is found whose first term is not divisible by the first term of the divisor, the exact division is impossible. In that case, write the last remainder after the quotient found, placing the divisor under it, in the form of a fraction.

SECOND EXAMPLE.

Let it be required to divide

$$51a^2b^2 + 10a^4 - 48a^3b - 15b^4 + 4ab^3 \text{ by } 4ab - 5a^2 + 3b^2.$$

We first arrange the dividend and divisor with reference to $a$.

$$\begin{array}{c}\text{\textit{Dividend.}} \qquad\qquad \text{\textit{Divisor.}}\end{array}$$

$$\dfrac{\begin{aligned}&10a^4- 48a^3b + 51a^2b^2+ 4ab^3-15b^4\\ +&10a^4-\ 8a^3b -\ 6a^2b^2\end{aligned}}{\begin{aligned}&-40a^3b + 57a^2b^2+ 4ab^3-15b^4\\ &-40a^3b + 32a^2b^2+24ab^3\end{aligned}}\;\Bigg|\;\begin{array}{l}-5a^2+ 4ab + 3b^2\\ \overline{-2a^2+ 8ab - 5b^2}\\ \text{\textit{Quotient.}}\end{array}$$

$$\begin{aligned}25a^2b^2-20ab^3-15b^4\\ 25a^2b^2-20ab^3-15b^4\end{aligned}$$

(3.)

$$\dfrac{\begin{aligned}&x^4 + x^3y + x^2y + xy^2 - 2y\\ &x^4 + x^3y\end{aligned}}{\begin{aligned}&\ + x^2y + xy^2\\ &\ + x^2y + xy^2\end{aligned}}\;\Bigg|\;\dfrac{x + y}{x^3+ xy}\qquad \dfrac{2y}{x+y}\,.$$

$$-\,2y$$

Here the division is not exact, and the quotient is fractional.

(4.)

$$\dfrac{\begin{aligned}1 &+ a\\ 1 &- a\end{aligned}}{\begin{aligned}&+ 2a\\ &+ 2a - 2a^2\end{aligned}}\;\Bigg|\;\begin{array}{l}1 - a\\ \overline{1 + 2a + 2a^2 + 2a^3 +},\ \&\text{c.}\end{array}$$

$$\dfrac{\begin{aligned}&+ 2a^2\\ &+ 2a^2 - 2a^3\end{aligned}}{\ + 2a^3}$$

In this example the operation does not terminate. It may be continued to any extent.

EXAMPLES.

1. Divide $a^2 + 2ax + x^2$ by $a + x$.  Ans. $a + x$.

2. Divide $a^3 - 3a^2y + 3ay^2 - y^3$ by $a - y$.

Ans. $a^2 - 2ay + y^2$.

4

3. Divide $24a^2b - 12a^3cb^2 - 6ab$ by $-6ab$.

$$\text{Ans. } +4a + 2a^2cb + 1.$$

4. Divide $6x^4 - 96$ by $3x - 6$.

$$\text{Ans. } 2x^3 + 4x^2 + 8x + 16.$$

5. Divide $a^5 - 5a^4x + 10a^3x^2 - 10a^2x^3 + 5ax^4 - x^5$ by $a^2 - 2ax + x^2$.     Ans. $a^3 - 3a^2x + 3ax^2 - x^3$.

6. Divide $48x^3 - 76ax^2 - 64a^2x + 105a^3$ by $2x - 3a$.

$$\text{Ans. } 24x^2 - 2ax - 35a^2.$$

$-$7. Divide $y^6 - 3y^4x^2 + 3y^2x^4 - x^6$ by $y^3 - 3y^2x + 3yx^2 - x^3$.     Ans. $y^3 + 3y^2x + 3yx^2 + x^3$.

8. Divide $64a^4b^6 - 25a^2b^8$ by $8a^2b^3 + 5ab^4$.

$$\text{Ans. } 8a^2b^3 - 5ab^4.$$

9. Divide $6a^3 + 23a^2b + 22ab^2 + 5b^3$ by $3a^2 + 4ab + b^2$.

$$\text{Ans. } 2a + 5b.$$

10. Divide $6ax^6 + 6ax^2y^6 + 42a^2x^2$ by $ax + 5ax$.

$$\text{Ans. } x^5 + xy^6 + 7ax.$$

11. Divide $-15a^4 + 37a^2bd - 29a^2cf - 20b^2d^2 + 44bcdf - 8c^2f^2$ by $3a^2 - 5bd + cf$.   Ans. $-5a^2 + 4bd - 8cf$.

12. Divide $x^4 + x^2y^2 + y^4$ by $x^2 - xy + y^2$.

$$\text{Ans. } x^2 + xy + y^2.$$

13. Divide $x^4 - y^4$ by $x - y$.

$$\text{Ans. } x^3 + x^2y + xy^2 + y^3.$$

14. Divide $3a^4 - 8a^2b^2 + 3a^2c^2 + 5b^4 - 3b^2c^2$ by $a^2 - b^2$.

$$\text{Ans. } 3a^2 - 5b^2 + 3c^2.$$

15. Divide $6x^6 - 5x^5y^2 - 6x^4y^4 + 6x^3y^2 + 15x^3y^3 - 9x^2y^4 + 10x^2y^5 + 15y^5$ by $3x^3 + 2x^2y^2 + 3y^2$.

$$\text{Ans. } 2x^3 - 3x^2y^2 + 5y^3.$$

16. Divide $-c^2 + 16a^2x^2 - 7abc - 4a^2bx - \blacksquare b^2 + 6acx$ by $8ax - 6ab - c$.     Ans. $2ax + ab + c$.

17. Divide $3x^4 + 4x^3y - 4x^2 - 4x^2y^2 + 16xy - 15$ by $2xy + x^2 - 3$.     Ans. $3x^2 - 2xy + 5$.

**18.** Divide $x^5 + 32y^5$ by $x + 2y$.

$Ans.$ $x^4 - 2x^3y + 4x^2y^2 - 8xy^3 + 16y^4$.

**19.** Divide $3a^4 - 26a^3b - 14ab^3 + 37a^2b^2$ by $2b^2 - 5ab + 3a^2$. $Ans.$ $a^2 - 7ab$.

**20.** Divide $a^4 - b^4$ by $a^3 + a^2b + ab^2 + b^3$.

$Ans.$ $a - b$.

**21.** Divide $x^3 - 3x^2y + y^3$ by $x + y$.

$Ans.$ $x^2 - 4xy + 4y^2 - \dfrac{3y^3}{x + y}$.

**22.** Divide $1 + 2a$ by $1 - a - a^2$.

$Ans.$ $1 + 3a + 4a^2 + 7a^3 +$, &c.

# CHAPTER III.

USEFU1 FORMULAS. FACTORING. GREATEST COMMON DIVISOR.
LEAST COMMON MULTIPLE.

## USEFUL FORMULAS.

**53.** A FORMULA is an algebraic expression of a general rule, or principle.

Formulas serve to shorten algebraic operations, and are also of much use in the operation of factoring. When translated into common language, they give rise to practical rules.

The verification of the following formulas affords additional exercises in Multiplication and Division.

### ( 1. )

**54.** To form the square of $a + b$, we have,

$$(a + b)^2 = (a + b)(a + b) = a^2 + 2ab + b^2.$$

That is,

*The square of the sum of any two quantities is equal to the square of the first, plus twice the product of the first by the second, plus the square of the second.*

1. Find the square of $2a + 3b$. We have from the rule,

$$(2a + 3b)^2 = 4a^2 + 12ab + 9b^2.$$

---

53. What is a formula? What are the uses of formulas?
54. What is the square of the sum of two quantities equal to?

2. Find the square of $5ab + 3ac$.

$Ans.$ $25a^2b^2 + 30a^2bc + 9a^2c^2$.

3. Find the square of $5a^2 + 8a^2b$.

$Ans.$ $25a^4 + 80a^4b + 64a^4b^2$.

4. Find the square of $6ax + 9a^2x^2$.

$Ans.$ $36a^2x^2 + 108a^3x^3 + 81a^4x^4$.

### (2.)

**55.** To form the square of a difference, $a - b$, we have,

$$(a - b)^2 = (a - b)(a - b) = a^2 - 2ab + b^2.$$

That is,

*The square of the difference of any two quantities is equal to the square of the first, minus twice the product of the first by the second, plus the square of the second.*

1. Find the square of $2a - b$. We have,

$$(2a - b)^2 = 4a^2 - 4ab + b^2.$$

2. Find the square of $4ac - bc$.

$Ans.$ $16a^2c^2 - 8abc^2 + b^2c^2$.

3. Find the square of $7a^2b^2 - 12ab^3$.

$Ans.$ $49a^4b^4 - 168a^3b^5 + 144a^2b^6$.

### (3.)

**56.** Multiply $a + b$ by $a - b$. We have,

$$(a + b) \times (a - b) = a^2 - b^2. \quad \text{Hence,}$$

*The sum of two quantities, multiplied by their difference, is equal to the difference of their squares.*

1. Multiply $2c + b$ by $2c - b$.     $Ans.$ $4c^2 - b^2$.

2. Multiply $9ac + 3bc$ by $9ac - 3bc$.

$Ans.$ $81a^2c^2 - 9b^2c^2$.

---

55. What is the square of the difference of two quantities equal to?

56. What is the sum of two quantities multiplied by their difference equal to?

**3.** Multiply $8a^3 + 7ab^2$ by $8a^3 - 7ab^2$.

$$Ans.\ 64a^6 - 49a^2b$$

**(4.)**

**57.** Multiply $a^2 + ab + b^2$ by $a - b$. We have,

$$(a^2 + ab + b^2)\ (a - b) = a^3 - b^3.$$

**(5.)**

**58.** Multiply $a^2 - ab + b^2$ by $a + b$. We have,

$$(a^2 - ab + b^2)\ (a + b) = a^3 + b^3.$$

**(6.)**

**59.** Multiply together, $a + b$, $a - b$, and $a^2 + b^2$. We have,

$$(a + b)\ (a - b)\ (a^2 + b^2) = a^4 - b^4.$$

**60.** Since every product is divisible by any of its factors, each formula establishes the principle set opposite its number.

**1.** *The sum of the squares of any two quantities, plus twice their product, is divisible by their sum.*

**2.** *The sum of the squares of any two quantities, minus twice their product, is divisible by the difference of the quantities.*

**3.** *The difference of the squares of any two quantities is divisible by the sum of the quantities, and also by their difference.*

**4.** *The difference of the cubes of any two quantities is divisible by the difference of the quantities ; also, by the sum of their squares, plus their product.*

- **5.** *The sum of the cubes of any two quantities is divisi-*

---

60. By what is any product divisible ?  By applying this principle, what follows from Formula (1)?  What from (2)?  What from (3)?  What from (4)?  What from (5)?  What from (6)?

*ble by the sum of the quantities ; also, by the sum of their squares minus their product.*

**6.** *The difference between the fourth powers of any two quantities is divisible by the sum of the quantities, by their difference, by the sum of their squares, and by the difference of their squares.*

---

## FACTORING.

**61.** Factoring is the operation of resolving a quantity into factors. The principles employed are the converse of those of Multiplication. The operations of factoring are performed by inspection.

1. What are the factors of the polynomial

$$ac + ab + ad.$$

We see, by inspection, that $a$ is a common factor of all the terms; hence, it may be placed without a parenthesis, and the other parts within; thus:

$$ac + ab + ad = a(c + b + d).$$

2. Find the factors of the polynomial $a^2b^2 + a^2d - a^2f$.
   Ans. $a^2(b^2 + d - f)$.

3. Find the factors of the polynomial $3a^2b - 6a^2b^2 + b^2d$.
   Ans. $b(3a^2 - 6a^2b + bd)$.

4. Find the factors of $3a^2b - 9a^2c - 18a^2xy$.
   Ans. $3a^2(b - 3c - 6xy)$.

5. Find the factors of $8a^2cx - 18acx^2 + 2ac^5y - 30a^6c^9$.
   Ans. $2ac(4ax - 9x^2 + c^4y - 15a^5c^8)$.

6. Factor $30a^4b^2c - 6a^3b^2d^3 + 18a^3b^2c^2$.
   Ans. $6a^3b^2(5ac - d^3 + 3c^2)$.

7. Factor $12c^4bd^3 - 15c^3d^4 - 6c^2d^3f$.
   Ans. $3c^2d^3(4c^2b - 5cd - 2f)$.

---

61. What is factoring?

80 ELEMENTARY ALGEBRA.

8. Factor $15a^3bcf - 10abc^4 - 25abcd$.

$$Ans.\ 5abc(3a^2f - 2c^3 - 5d).$$

**62.** When two terms of a trinomial are squares, and positive, and the third term is equal to twice the product of their square roots, the trinomial may be resolved into factors by Formula ( **1** ).

1. Factor $a^2 + 2ab + b^2$.    $Ans.\ (a + b)(a + b)$.

2. Factor $4a^2 + 12ab + 9b^2$.
$$Ans.\ (2a + 3b)(2a + 3b).$$

3. Factor $9a^2 + 12ab + 4b^2$.
$$Ans.\ (3a + 2b)(3a + 2b).$$

4. Factor $4x^2 + 8x + 4$.    $Ans.\ (2x + 2)(2x + 2)$.

5. Factor $9a^2b^2 + 12abc + 4c^2$.
$$Ans.\ (3ab + 2c)(3ab + 2c).$$

6. Factor $16x^2y^2 + 16xy^3 + 4y^4$.
$$Ans.\ (4xy + 2y^2)(4xy + 2y^2).$$

**63.** When two terms of a trinomial are squares, and positive, and the third term is equal to *minus* twice their square roots, the trinomial may be factored by Formula ( **2** ).

1. Factor $a^2 - 2ab + b^2$.    $Ans.\ (a - b)(a - b)$.

2. Factor $4a^2 - 4ab + b^2$.    $Ans.\ (2a - b)(2a - b)$.

3. Factor $9a^2 - 6ac + c^2$.    $Ans.\ (3a - c)(3a - c)$.

4. Factor $a^2x^2 - 4ax + 4$.    $Ans.\ (ax - 2)(ax - 2)$.

5. Factor $4x^2 - 4xy + y^2$.    $Ans.\ (2x - y)(2x - y)$.

**6.** Factor $36x^2 - 24xy + 4y^2$.

$Ans.$ $(6x - 2y)(6x - 2y)$.

**64.** When the two terms of a binomial are squares and have contrary signs, the binomial may be factored by Formula (**3**).

1. Factor $4c^2 - b^2$.    $Ans.$ $(2c + b)(2c - b)$

2. Factor $81a^2c^2 - 9b^2c^2$.

$Ans.$ $(9ac + 3bc)(9ac - 3bc)$.

3. Factor $64a^4b^4 - 25x^2y^2$.

$Ans.$ $(8a^2b^2 + 5xy)(8a^2b^2 - 5xy)$.

4. Factor $25a^2c^2 - 9x^4y^2$.

$Ans.$ $(5ac + 3x^2y)(5ac - 3x^2y)$.

5. Factor $36a^4b^4c^2 - 9x^6$.

$Ans.$ $(6a^2b^2c + 3x^3)(6a^2b^2c - 3x^3)$.

6. Factor $49x^4 - 36y^4$. $Ans.$ $(7x^2 + 6y^2)(7x^2 - 6y^2)$.

**65.** When the two terms of a binomial are cubes, and have contrary signs, the binomial may be factored by Formula (**4**).

1. Factor $8a^3 - c^3$.    $Ans.$ $(2a - c)(4a^2 + 2ac + c^2)$.

2. Factor $27a^3 - 64$.

$Ans.$ $(3a - 4)(9a^2 + 12a + 16)$.

3. Factor $a^3 - 64b^3$.

$Ans.$ $(a - 4b)(a^2 + 4ab + 16b^2)$.

4. Factor $a^3 - .27b^3$. $Ans.$ $(a - 3b)(a^2 + 3ab + 9b^2)$.

**66.** When the terms of a binomial are cubes and have like signs, the binomial may be factored by Formula ( **5** ).

1. Factor  $8a^3 + c^3$ .   *Ans.*  $(2a + c)(4a^2 - 2ac + c^2)$ .

2. Factor  $27a^3 + 64$ .
$$Ans. \ (3a + 4)(9a^2 - 12a + 16).$$
3. Factor  $a^3 + 64b^3$ .
$$Ans. \ (a + 4b)(a^2 - 4ab + 16b^2).$$

4. Factor  $a^3 + 27b^3$ .   *Ans.*  $(a + 3b)(a^2 - 3ab + 9b^2)$ .

**67.** When the terms of a binomial are 4th powers, and have contrary signs, the binomial may be factored by Formula ( **6** ).

1. What are the factors of  $a^4 - b^4$ ?
$$Ans. \ (a + b)(a - b)(a^2 + b^2).$$

2. What are the factors of  $81a^4 - 16b^4$ ?
$$Ans. \ (3a + 2b)(3a - 2b)(9a^2 + 4b^2).$$

3. What are the factors of  $16a^4b^4 - 81c^4d^4$ ?
$$Ans. \ (2ab + 3cd)(2ab - 3cd)(4a^2b^2 + 9c^2d^2).$$

### GREATEST COMMON DIVISOR.

**68.** A COMMON DIVISOR of two quantities, is a quantity that will divide them both without a remainder. Thus,  $3a^2b$ , is a common divisor of  $9a^2b^2c$  and  $3a^2b^2 - 6a^3b^3$ .

---

66. When may a binomial be factored by this method?
67. When may a binomial be factored by this method?
68. What is the common divisor of two quantities?

**69.** A SIMPLE or PRIME FACTOR is one that cannot be resolved into any other factors.

Every prime factor, common to two quantities, is a common divisor of those quantities. The continued product of any number of prime factors, common to two quantities, is also a common divisor of those quantities.

**70.** The GREATEST COMMON DIVISOR of two quantities, is the continued product of all the prime factors which are common to both.

**71.** When both quantities can be resolved into prime factors, by the method of factoring already given, the greatest common divisor may be found by the following

RULE.

I. *Resolve both quantities into their prime factors :*

II. *Find the continued product of all the factors which are common to both ; it will be the greatest common divisor required.*

EXAMPLES.

1. Required the greatest common divisor of $75a^2b^2c$ and $25abd$. Factoring, we have,

$$75a^2b^2c = 3 \times 5 \times 5aabbc$$
$$25abd = 5 \times 5abd.$$

The factors, 5, 5, $a$ and $b$, are common; hence,

$$5 \times 5 \times a \times b = 25ab,$$

is the divisor sought.

---

69. What is a simple or prime factor? Is a prime factor, common to two quantities, a common divisor?

70. What is the greatest common divisor?

71. If both quantities can be resolved into prime factors, how do you find the greatest common divisor?

### VERIFICATION.

$$75a^2b^2c \div 25ab = 3abc$$
$$25abd \div 25ab = d;$$

and since the quotients have no common factor, they cannot be further divided.

2. Required the greatest common divisor of $a^2 - 2ab + b^2$ and $a^2 - b^2$.       *Ans. a - b.*

3. Required the greatest common divisor of $a^2 + 2ab + b^2$ and $a + b$.       *Ans. a + b.*

4. Required the greatest common divisor of $a^2x^2 - 4ax + 4$ and $ax - 2$.       *Ans. ax - 2.*

5. Find the greatest common divisor of $3a^2b - 9a^2c - 18a^2xy$ and $b^2c - 3bc^2 - 6bcxy$.   *Ans. b - 3c - 6xy.*

6. Find the greatest common divisor of $4a^2c - 4acx$ and $3a^2g - 3agx$.       *Ans. a(a - x),* or $a^2 - ax$.

7. Find the greatest common divisor of $4c^2 - 12cx + 9x^2$ and $4c^2 - 9x^2$.       *Ans. 2c - 3x.*

8. Find the greatest common divisor of $x^3 - y^3$ and $x^2 - y^2$.       *Ans. x - y.*

9. Find the greatest common divisor of $4c^2 + 4bc + b^2$ and $4c^2 - b^2$.       *Ans. 2c + b.*

10. Find the greatest common divisor of $25a^2c^2 - 9x^4y^4$ and $5acd^2 + 3d^2x^2y^2$.       *Ans. 5ac + 3x^2y^2.*

NOTE.—To find the greatest common divisor of three quantities. First find the greatest common divisor of two of them, and then the greatest common divisor between this result and the third.

1. What is the greatest common divisor of $4ax^2y$, $16abx^2$, and $24acx^2$?       *Ans. 4ax^2.*

2. Of $3x^2 - 6x$, $2x^3 - 4x^2$, and $x^2y - 2xy$?   *Ans. x^2 - 2x.*

---

72. When is one quantity a multiple of another?

## LEAST COMMON MULTIPLE.

**72.** One quantity is a MULTIPLE of another, when it can be divided by that other without a remainder. Thus, $8a^2b$, is a multiple of 8, also of $a^2$, and of $b$. *

**73.** A quantity is a *Common Multiple* of two or more quantities, when it can be divided by each, separately, without a remainder. Thus, $24a^3x^3$, is a common multiple of $6ax$ and $4a^2x$.

**74.** The LEAST COMMON MULTIPLE of two or more quantities, is the simplest quantity that can be divided by each, without a remainder. Thus, $12a^2b^2x^2$, is the least common multiple of $2a^2x$, $4ab^2$, and $6a^2b^2x^2$.

**75.** Since the common multiple is a dividend of each of the quantities, and since the division is exact, the common multiple must contain every prime factor in all the quantities; and if the same factor enters more than once, it must enter an equal number of times into the common multiple.

When the given quantities can be factored, by any of the methods already given, the least common multiple may be found by the following

### RULE.

I. *Resolve each of the quantities into its prime factors.*

II. *Take each factor as many times as it enters any one of the quantities, and form the continued product of these factors ; it will be the least common multiple.*

---

73. When is a quantity a common multiple of several others?

74. What is the least common multiple of two or more quantities?

75. What does the common multiple of two or more quantities contain, as factors? How may the least common multiple be found?

* The *multiple* of a quantity, is simply a *dividend* which will give an exact quotient.

1. Find the least common multiple of $12a^3b^2c^2$ and $8a^2b^3$.

$$12a^3b^2c^2 = 2.2.3.aaabbcc.$$
$$8a^2b^3 = 2.2.2.aabbb.$$

Now, since 2 enters 3 times as a factor, it must enter 3 times in the common multiple: 3 must enter once; $a$, 3 times; $b$, 3 times; and $c$, twice; hence,

$$2.2.2.3aaabbbcc = 24a^3b^3c^2,$$

is the least common multiple.

Find the least common multiples of the following:

2. $6a$, $5a^2b$, and $25abc^2$.      *Ans.* $150a^2bc^2$.

3. $3a^2b$, $9abc$, and $27a^2x^2$.      *Ans.* $27a^2bcx^2$.

4. $4a^2x^2y^2$, $8a^3xy$, $16a^4y^3$, and $24a^5y^4x$.   *Ans.* $48a^5x^2y^4$.

5. $ax - bx$, $ay - by$, and $x^2y^2$.

       *Ans.* $(a - b)x.x.yy = ax^2y^2 - bx^2y^2$.

6. $a + b$, $a^2 - b^2$, and $a^2 + 2ab + b^2$.

       *Ans.* $(a + b)^2 (a - b)$.

7. $3a^3b^2$, $9a^2x^2$, $18a^4y^3$, $3a^2y^2$.    *Ans.* $18a^4b^2x^2y^3$.

8. $8a^2(a - b)$, $15a^5(a - b)^2$, and $12a^3(a^2 - b^2)$.

       *Ans.* $120a^5(a - b)^2 (a + b)$.

●

# CHAPTER IV.

## FRACTIONS.

**76.** If the unit 1 be divided into any number of equal parts, each part is called a FRACTIONAL UNIT. Thus, $\frac{1}{2}$, $\frac{1}{4}$, $\frac{1}{7}$, $\frac{1}{b}$, are fractional units.

**77.** A FRACTION is a fractional unit, or a collection of fractional units. Thus, $\frac{1}{2}$, $\frac{3}{4}$, $\frac{5}{7}$, $\frac{a}{b}$, are fractions.

**78.** Every fraction is composed of two parts, the De-nominator and Numerator. The *Denominator* shows into how many equal parts the unit 1 is divided; and the *Nu-merator* how many of these parts are taken. Thus, in the fraction $\frac{a}{b}$, the denominator $b$, shows that 1 is divided into $b$ equal parts, and the numerator $a$, shows that $a$ of these parts are taken. The fractional unit, in all cases, is equal to the reciprocal of the denominator.

---

76. If 1 be divided into any number of equal parts, what is each part called?

77. What is a fraction?

78. Of how many parts is any fraction composed? What are they called? What does the denominator show? What the numerator? What is the fractional unit equal to?

**79.** An Entire Quantity is one which contains no fractional part. Thus, 7, 11, $a^3x$, $4x^2 - 3y$, are entire quantities.

An entire quantity may be regarded as a fraction whose denominator is 1. Thus, $7 = \dfrac{7}{1}$, $ab = \dfrac{ab}{1}$.

**80.** A Mixed Quantity is a quantity containing both entire and fractional parts. Thus, $7\frac{4}{10}$, $8\frac{3}{7}$, $a + \dfrac{bx}{c}$, are mixed quantities.

**81.** Let $\dfrac{a}{b}$ denote any fraction, and $q$ any quantity whatever. From the preceding definitions, $\dfrac{a}{b}$ denotes that $\dfrac{1}{b}$ is taken $a$ times; also, $\dfrac{aq}{b}$ denotes that $\dfrac{1}{b}$ is taken $aq$ times; that is,

$$\frac{aq}{b} = \frac{a}{b} \times q; \text{ hence,}$$

*Multiplying the numerator of a fraction by any quantity, is equivalent to multiplying the fraction by that quantity.*

We see, also, that *any quantity may be multiplied by a fraction, by multiplying it by the numerator, and then dividing the result by the denominator.*

**82.** It is a principle of Division, that the same result will be obtained if we divide the quantity $a$ by the product of two factors, $p \times q$, as would be obtained by dividing it

79. What is an entire quantity?  When may it be regarded as a frac tion?

80. What is a mixed quantity?

81. How may a fraction be multiplied by any quantity?

82 How may a fraction be divided by any quantity?

first by one of the factors, $p$, and then dividing that result by the other factor, $q$. That is,

$$\frac{a}{pq} = \left(\frac{a}{p}\right) \div q; \quad \text{or,} \quad \frac{a}{pq} = \left(\frac{a}{q}\right) \div p; \text{ hence,}$$

*Multiplying the denominator of a fraction by any quantity, is equivalent to dividing the fraction by that quantity.*

**83.** Since the operations of Multiplication and Division are the converse of each other, it follows, from the preceding principles, that,

*Dividing the numerator of a fraction by any quantity, is equivalent to dividing the fraction by that quantity;* and,

*Dividing the denominator of a fraction by any quantity, is equivalent to multiplying the fraction by that quantity.*

**84.** Since a quantity may be multiplied, and the result divided by the same quantity, without altering the value, it follows that,

*Both terms of a fraction may be multiplied by any quantity, or both divided by any quantity, without changing the value of the fraction.*

---

### TRANSFORMATION OF FRACTIONS.

**85.** The *transformation* of a quantity, is the operation of changing its form, without altering its value. The term *reduce* has a technical signification, and means, to *Transform*.

---

83. What follows from the preceding principles?

84. What operations may be performed without altering the value of a fraction?

85. What is the transformation of a quantity?

### FIRST TRANSFORMATION.

*To reduce an entire quantity to a fractional form having a given denominator.*

**86.** Let $a$ be the quantity, and $b$ the given denominator. We have, evidently, $a = \dfrac{ab}{b}$; hence, the

### RULE.

*Multiply the quantity by the given denominator, and write the product over this given denominator.*

### SECOND TRANSFORMATION.

*To reduce a fraction to its lowest terms.*

**87.** ' A fraction is in its *lowest terms*, when the numerator and denominator contain no common factors.

It has been shown, that both terms of a fraction may be divided by the same quantity, without altering its value. Hence, if they have any common factors, we may strike them out.

### RULE.

*Resolve each term of the fraction into its prime factors; then strike out all that are common to both.*

The same result is attained by dividing both terms of the fraction by any quantity that will divide them, without a remainder; or, by dividing them by their greatest common divisor.

---

86. How do you reduce an entire quantity to a fractional form having a given denominator?

87. How do you reduce a fraction to its lowest terms?

1. Reduce $\dfrac{15a^2c^2}{25acd}$ to its lowest terms.

Factoring,    $\dfrac{15a^2c^2}{25acd} = \dfrac{3.5aacc}{5.5acd}$;

Canceling the common factors, 5, $a$, and $c$, we have,

$$\dfrac{15a^2c^2}{25acd} = \dfrac{3ac}{5d} \cdot \quad Ans.$$

2. Reduce $\dfrac{85b^7cd^5}{15b^7c^8d^5}$.    $Ans.$ $\dfrac{17}{3c^7}$.

3. Reduce $\dfrac{60c^6d^4f^5}{12c^5d^8f^9}$.    $Ans.$ $\dfrac{5c}{d^4f^4}$.

4. Reduce $\dfrac{ab - ac}{b - c}$.    $Ans.$ $\dfrac{a}{1} = a$.

5. Reduce $\dfrac{n^2 - 2n + 1}{n^2 - 1}$.    $Ans.$ $\dfrac{n - 1}{n + 1}$.

6. Reduce $\dfrac{x^3 - ax^2}{x^2 - 2ax + a^2}$.    $Ans.$ $\dfrac{x^2}{x - a}$.

7. Reduce $\dfrac{96a^3b^2c}{- 12a^3b^2c}$.    $Ans.$ $-\dfrac{8}{1} = -8$.

8. Reduce $\dfrac{24b^5 - 36ab^4}{48a^4b^4 - 66a^5b^6}$.    $Ans.$ $\dfrac{4b - 6a}{8a^4 - 11a^5b^2}$.

9. Reduce $\dfrac{a^2 - b^2}{a^2 - 2ab + b^2}$.    $Ans.$ $\dfrac{a + b}{a - b}$.

10. Reduce $\dfrac{5a^3 - 10a^2b + 5ab^2}{8a^3 - 8a^2b}$.    $Ans.$ $\dfrac{5(a - b)}{8a}$.

11. Reduce $\dfrac{3a^2 + 6a^2b^2}{12a^4 + 6a^3c^2}$.    $Ans.$ $\dfrac{1 + 2b^2}{4a^2 + 2ac^2}$.

12. Reduce $\dfrac{a^2 + 2ax + x^2}{3(a^2 - x^2)}$.    $Ans.$ $\dfrac{a + x}{3(a - x)}$.

### THIRD TRANSFORMATION.

*To reduce a fraction to a mixed quantity.*

**88.** When any term of the numerator is divisible by any term of the denominator, the transformation can be effected by Division.

### RULE.

*Perform the indicated division, continuing the operation as far as possible ; then write the remainder over the denominator, and annex the result to the quotient found.*

### EXAMPLES.

1. Reduce $\dfrac{ax - a^2}{x}$.  *Ans.* $a - \dfrac{a^2}{x}$.

2. Reduce $\dfrac{ax - x^2}{x}$.  *Ans.* $a - x$.

3. Reduce $\dfrac{ab - 2a^2}{b}$.  *Ans.* $a - \dfrac{2a^2}{b}$.

4. Reduce $\dfrac{a^2 - x^2}{a - x}$.  *Ans.* $a + x$.

5. Reduce $\dfrac{x^3 - y^3}{x - y}$.  *Ans.* $x^2 + xy + y^2$.

6. Reduce $\dfrac{10x^2 - 5x + 3}{5x}$.  *Ans.* $2x - 1 + \dfrac{3}{5x}$.

7. Reduce $\dfrac{36x^3 - 72x + 32a^2x^2}{9x}$ .. $4x^2 - 8 + \dfrac{32a^2x}{9}$.

8. Reduce $\dfrac{18acf - 6bdcf - 2ad}{3adf}$ .. $\dfrac{6c}{d} - \dfrac{2bc}{a} - \dfrac{2}{3f}$.

9. Reduce $\dfrac{x^2 + x - 4}{x + 2}$.  *Ans.* $x - 1 - \dfrac{2}{x + 2}$.

---

**88.** How do you reduce a fraction to a mixed quantity ?

10. Reduce $\dfrac{a^2 + b^2}{a + b}$.  $Ans.$ $a - b + \dfrac{2b^2}{a + b}$.

11. Reduce $\dfrac{x^2 + 3x - 25}{x - 4}$.  $Ans.$ $x + 7 + \dfrac{3}{x - 4}$.

### FOURTH TRANSFORMATION.

*To reduce a mixed quantity to a fractional form.*

**89.** This transformation is the converse of the preceding, and may be effected by the following

### RULE.

*Multiply the entire part by the denominator of the fraction, and add to the product the numerator; write the result over the denominator of the fraction.*

### EXAMPLES.

1. Reduce $6\frac{1}{7}$ to the form of a fraction.

$6 \times 7 = 42$; $42 + 1 = 43$; hence, $6\frac{1}{7} = \dfrac{43}{7}$.

Reduce the following to fractional forms:

2. $x - \dfrac{a^2 - x^2}{x} = \dfrac{x^2 - (a^2 - x^2)}{x}$.  $Ans.$ $\dfrac{2x^2 - a^2}{x}$.

3. $x - \dfrac{ax + x^2}{2a}$.  $Ans.$ $\dfrac{ax - x^2}{2a}$.

4. $5 + \dfrac{2x - 7}{3x}$.  $Ans.$ $\dfrac{17x - 7}{3x}$.

5. $1 - \dfrac{x - a - 1}{a}$.  $Ans.$ $\dfrac{2a - x + 1}{a}$.

6. $1 + 2x - \dfrac{x - 3}{5x}$.  $Ans.$ $\dfrac{10x^2 + 4x + 3}{5x}$.

89. How do you reduce a mixed quantity to a fractional form?

7. $2a + b - \dfrac{3c + 4}{8}$.     *Ans.* $\dfrac{16a + 8b - 3c - 4}{8}$.

8. $6ax + b - \dfrac{6a^2x - ab}{4a}$.     *Ans.* $\dfrac{18a^2x + 5ab}{4a}$.

9. $8 + 3ab - \dfrac{8 + 6a^2b^2x^4}{12abx^4}$.

*Ans.* $\dfrac{96abx^4 + 30a^2b^2x^4 - 8}{12abx^4}$.

### FIFTH TRANSFORMATION.

*To reduce fractions having different denominators, to equivalent fractions having the least common denominator.*

**90.** This transformation is effected by finding the least common multiple of the denominators.

1. Reduce $\dfrac{1}{3}$, $\dfrac{3}{4}$, and $\dfrac{5}{12}$, to their least common denominators.

The least common multiple of the denominators is 12, which is also the least common denominator of the required fractions. If each fraction be multiplied by 12, and the result divided by 12, the values of the fractions will not be changed.

$\dfrac{1}{3} \times 12 = 4$,  1st new numerator;

$\dfrac{3}{4} \times 12 = 9$,  2d new numerator;

$\dfrac{5}{12} \times 12 = 5$,  3rd new numerator; hence,

$\dfrac{4}{12}$, $\dfrac{9}{12}$, and $\dfrac{5}{12}$ are the new equivalent fractions.

---

90. How do you reduce fractions having different denominators, to equivalent fractions having the least common denominator? When the numerators have no common factor, how do you reduce them?

I. *Find the least common multiple of the denominators:*

II. *Multiply each fraction by it, and cancel the denominator:*

III. *Write each product over the common multiple, and the results will be the required fractions.*

### GENERAL RULE.

*Multiply each numerator by all the denominators except its own, for the new numerators, and all the denominators together for a common denominator.*

### EXAMPLES.

1. Reduce $\dfrac{a}{a^2 - b^2}$ and $\dfrac{c}{a + b}$ to their least common denominator.

The least common multiple of the denominators is $(a + b)(a - b)$:

$$\frac{a}{a^2 - b^2} \times (a + b)(a - b) = a$$

$$\frac{c}{a + b} \times (a + b)(a - b) = c(a - b; \text{ hence,}$$

$\dfrac{a}{(a + b)(a - b)}$ and $\dfrac{c(a - b)}{(a + b)(a - b)}$, are the required fractions.

Reduce the following to their least common denominators:

2. $\dfrac{3x}{4}, \dfrac{4}{6},$ and $\dfrac{12x^2}{15}$.    *Ans.* $\dfrac{45x}{60}, \dfrac{40}{60}, \dfrac{48x^2}{60}$.

3. $a, \dfrac{3b^2}{4},$ and $\dfrac{5c^3}{6}$.    *Ans.* $\dfrac{12a}{12}, \dfrac{9b^2}{12}, \dfrac{10c^3}{12}$.

4. $\dfrac{3x}{2a}, \dfrac{2b}{3c},$ and $d$.    *Ans.* $\dfrac{9cx}{6ac}, \dfrac{4ab}{6ac}, \dfrac{6acd}{6ac}$.

5. $\dfrac{3}{4}$, $\dfrac{2x}{3}$, $a + \dfrac{2x}{a}$.   $Ans.$   $\dfrac{9a}{12a}$, $\dfrac{8ax}{12a}$, $\dfrac{12a^2 + 24x}{12a}$.

6. $\dfrac{x}{1 - x}$, $\dfrac{x^2}{(1 - x)^2}$, and $\dfrac{x^3}{(1 - x)^3}$.

$Ans.$   $\dfrac{x(1 - x)^2}{(1 - x)^3}$, $\dfrac{x^2(1 - x)}{(1 - x)^3}$, and $\dfrac{x^3}{(1 - x)^3}$.

7. $\dfrac{c}{5a}$, $\dfrac{c - b}{c}$, and $\dfrac{c}{a + b}$.

$\dfrac{ac^2 + bc^2}{5a^2c + 5abc}$, $\dfrac{5a^2c - 5a^2b + 5abc - 5ab^2}{5a^2c + 5abc}$, $\dfrac{5ac^2}{5a^2c + 5abc}$.

8. $\dfrac{cx}{a - x}$, $\dfrac{dx^2}{a + x}$, and $\dfrac{x^3}{a + x}$.

$Ans.$   $\dfrac{cx(a + x)}{a^2 - x^2}$, $\dfrac{dx^2(a - x)}{a^2 - x^2}$, and $\dfrac{x^3(a - x)}{a^2 - x^2}$.

---

### ADDITION OF FRACTIONS.

**91.** Fractions can only be added when they have a common unit, that is, when they have a common denominator. In that case, the sum of the numerators will indicate how many times that unit is taken in the entire collection. Hence, the

### RULE.

I. *Reduce the fractions to be added, to a common denominator :*

II. *Add the numerators together for a new numerator, and write the sum over the common denominator.*

### EXAMPLES.

1. Add $\dfrac{6}{2}$, $\dfrac{4}{3}$, and $\dfrac{2}{5}$, together.

---

91. What is the rule for adding fractions?

By reducing to a common denominator, we have,

$$6 \times 3 \times 5 = 90, \quad \text{1st numerator.}$$
$$4 \times 2 \times 5 = 40, \quad \text{2d numerator.}$$
$$2 \times 3 \times 2 = 12, \quad \text{3d numerator.}$$
$$2 \times 3 \times 5 = 30, \quad \text{the denominator.}$$

Hence, the expression for the sum of the fractions becomes

$$\frac{90}{30} + \frac{40}{30} + \frac{12}{30} = \frac{142}{30};$$

which, being reduced to the simplest form, gives $4\frac{11}{15}$.

2. Find the sum of $\frac{a}{b}$, $\frac{c}{d}$, and $\frac{e}{f}$.

Here, $\left. \begin{array}{l} a \times d \times f = adf \\ c \times b \times f = cbf \\ e \times b \times d = ebd \end{array} \right\}$ the new numerators.

and $\quad b \times d \times f = bdf \quad$ the common denominator.

Hence, $\dfrac{adf}{bdf} + \dfrac{cbf}{bdf} + \dfrac{ebd}{bdf} = \dfrac{adf + cbf + ebd}{bdf}$, the sum.

Add the following:

3. $a - \dfrac{3x^2}{b}$, and $b + \dfrac{2ax}{c}$. Ans. $a + b + \dfrac{2abx - 3cx^2}{bc}$.

4. $\dfrac{x}{2}$, $\dfrac{x}{3}$, and $\dfrac{x}{4}$. Ans. $x + \dfrac{x}{12}$.

5. $\dfrac{x - 2}{3}$ and $\dfrac{4x}{7}$. Ans. $\dfrac{19x - 14}{21}$.

6. $x + \dfrac{x - 2}{3}$ and $3x + \dfrac{2x - 3}{4}$. Ans: $4x + \dfrac{10x - 17}{12}$.

7. $4x$, $\dfrac{5x^2}{2a}$, and $\dfrac{x + a}{2x}$. Ans. $4x + \dfrac{5x^3 + ax + a^2}{2ax}$.

8. $\dfrac{2x}{3}$, $\dfrac{7x}{4}$, and $\dfrac{2x + 1}{5}$. Ans. $2x + \dfrac{49x + 12}{60}$.

9. $4x$, $\dfrac{7x}{9}$, and $2 + \dfrac{x}{5}$. Ans. $2 + 4x + \dfrac{44x}{45}$.

5

10. $3x + \dfrac{2x}{5}$ and $x - \dfrac{8x}{9}$.　　　　 *Ans.* $3x + \dfrac{23x}{45}$

11. $ac - \dfrac{6b}{8a}$ and $1 - \dfrac{c}{d}$.

　　　　　　　　 *Ans.* $1 + ac - \dfrac{6bd + 8ac}{8ad}$.

12. $\dfrac{3}{(x-1)^3}$, $\dfrac{3}{(x-1)^2}$, and $\dfrac{4}{x-1}$.

　　　　　　　　　 *Ans.* $\dfrac{4x^2 - 5x + 4}{(x-1)^3}$.

13. $\dfrac{1}{4(1+a)}$, $\dfrac{1}{4(1-a)}$, and $\dfrac{1}{2(1-a^2)}$. 　*Ans.* $\dfrac{1}{1-a^2}$.

## SUBTRACTION OF FRACTIONS.

**92.** Fractions can only be subtracted when they have the same unit; that is, a common denominator. In that case, the numerator of the minuend, *minus* that of the subtrahend, will indicate the number of times that the common unit is to be taken in the difference. Hence, the

### RULE.

I. *Reduce the two fractions to a common denominator:*

II. *Then subtract the numerator of the subtrahend from that of the minuend for a new numerator, and write the remainder over the common denominator.*

### EXAMPLES.

1. What is the difference between $\dfrac{3}{7}$ and $\dfrac{2}{8}$.

$$\frac{3}{7} - \frac{2}{8} = \frac{24}{56} - \frac{14}{56} = \frac{10}{56} = \frac{5}{28}\text{, } Ans.$$

---

92. What is the rule for subtracting fractions?

2. Find the difference of the fractions $\dfrac{x-a}{2b}$ and $\dfrac{2a-4x}{3c}$.

Here, $\left\{ \begin{array}{l} (x-a) \times 3c = 3cx - 3ac \\ (2a - 4x) \times 2b = 4ab - 8bx \end{array} \right\}$ the numerators,

and, $\qquad 2b \times 3c = 6bc$ the common denominator.

Hence, $\dfrac{3cx-3ac}{6bc} - \dfrac{4ab-8bx}{6bc} = \dfrac{3cx-3ac-4ab+8bx}{6bc}$. *Ans.*

3. Required the difference of $\dfrac{12x}{7}$ and $\dfrac{3x}{5}$.    *Ans.* $\dfrac{39x}{35}$.

4. Required the difference of $5y$ and $\dfrac{3y}{8}$.    *Ans.* $\dfrac{37y}{8}$.

5. Required the difference of $\dfrac{3x}{7}$ and $\dfrac{2x}{9}$.    *Ans.* $\dfrac{13x}{63}$.

6. From $\dfrac{x+y}{x-y}$ subtract $\dfrac{x-y}{x+y}$.    *Ans.* $\dfrac{4xy}{x^2-y^2}$.

7. From $\dfrac{1}{y-z}$ subtract $\dfrac{1}{y^2-z^2}$.    *Ans.* $\dfrac{y+z-1}{y^2-z^2}$.

Find the differences of the following:

8. $\dfrac{3x+a}{5b}$ and $\dfrac{2x+7}{8}$.   *Ans.* $\dfrac{24x+8a-10bx-35b}{40b}$.

9. $3x + \dfrac{x}{b}$ and $x - \dfrac{x-a}{c}$.   *Ans.* $2x + \dfrac{cx+bx-ab}{bc}$.

10. $a + \dfrac{a-x}{a(a+x)}$ and $\dfrac{a+x}{a(a-x)}$.   *Ans.* $a - \dfrac{4x}{a^2-x^2}$.

---

## MULTIPLICATION OF FRACTIONS.

**93.** Let $\dfrac{a}{b}$ and $\dfrac{c}{d}$, represent any two fractions. It has been shown (Art. 81), that any quantity may be multiplied

---

93. What is the rule for the multiplication of fractions?

by a fraction, by first multiplying by the numerator, and then dividing the result by the denominator.

To multiply $\frac{a}{b}$ by $\frac{c}{d}$, we first multiply by $c$, giving $\frac{ac}{b}$; then, we divide this result by $d$, which is done by multiplying the denominator by $d$; this gives for the product, $\frac{ac}{bd}$; that is,

$$\frac{a}{b} \times \frac{c}{d} = \frac{ac}{bd}; \text{ hence,}$$

### RULE.

I. *If there are mixed quantities, reduce them to a fractional form; then,*

II. *Multiply the numerators together for a new numerator, and the denominators for a new denominator.*

### EXAMPLES.

1. Multiply $a + \frac{bx}{a}$ by $\frac{c}{d}$. First, $a + \frac{bx}{a} = \frac{a^2 + bx}{a}$,

hence,    $\frac{a^2 + bx}{a} \times \frac{c}{d} = \frac{a^2c + bcx}{ad}$. *Ans.*

Find the products of the following quantities:

2. $\frac{2x}{a}$, $\frac{3ab}{c}$, and $\frac{3ac}{2b}$    *Ans. 9ax.*

3. $b + \frac{bx}{a}$ and $\frac{a}{x}$.    *Ans.* $\frac{ab + bx}{x}$.

4. $\frac{a^2 - b^2}{bc}$ and $\frac{x^2 + b^2}{b + c}$.    *Ans.* $\frac{x^4 - b^4}{b^2c + bc^2}$.

5. $x + \frac{x+1}{a}$, and $\frac{x-1}{a+b}$.    *Ans.* $\frac{ax^2 - ax + x^2 - 1}{a^2 + ab}$.

6. $a + \frac{ax}{a-x}$ and $\frac{a^2 - x^2}{x + x^2}$.    *Ans.* $\frac{a^3 + a^2x}{x + x^2}$.

**7.** Multiply $\dfrac{2a}{a-b}$ by $\dfrac{a^2-b^2}{3}$.

We have, by the rule,

$$\dfrac{2a}{a-b} \times \dfrac{a^2-b^2}{3} = \dfrac{2a(a^2-b^2)}{3(a-b)} = \dfrac{2a(a+b)(a-b)}{3(a-b)}$$
$$= \dfrac{2a}{3}(a+b).$$

After indicating the operation, we factored both numerator and denominator, and then canceled the common factors, before performing the multiplication. *This should be done, whenever there are common factors.*

**8.** $\dfrac{2}{x-y}$ by $\dfrac{x^2-y^2}{a}$.      *Ans.* $\dfrac{2(x+y)}{a}$.

**9.** $\dfrac{x^2-4}{3}$ by $\dfrac{4x}{x+2}$.      *Ans.* $\dfrac{4x(x-2)}{3}$.

**10.** $\dfrac{(a+b)^2}{2x}$ by $\dfrac{4x^2}{(a+b)}$.      *Ans.* $2x(a+b)$.

**11.** $\dfrac{(x-1)^2}{y^3}$ by $\dfrac{(x+1)y^2}{x-1}$.      *Ans.* $\dfrac{x^2-1}{y}$.

**12.** $\dfrac{(a^2-x^2)}{1-x^2}$ by $\dfrac{1+x}{a+x}$.      *Ans.* $\dfrac{a-x}{1-x}$.

**13.** $x+\dfrac{2xy}{x-y}$ by $x-\dfrac{2xy}{x+y}$      *Ans.* $x^2$.

**14.** $\dfrac{2a-b}{4a}$ by $\dfrac{6a-2b}{b^2-2ab}$.      *Ans.* $\dfrac{b-3a}{2ab}$.

**15.** $x-\dfrac{y^2}{x}$ by $\dfrac{x}{y}+\dfrac{y}{x}$.      *Ans.* $\dfrac{x^4-y^4}{x^2y}$.

## DIVISION OF FRACTIONS.

**94.** Since $\dfrac{p}{q} = p \times \dfrac{1}{q}$, it follows that, dividing by a quantity is equivalent to multiplying by its reciprocal. But the reciprocal of a fraction, $\dfrac{c}{d}$, is $\dfrac{d}{c}$ (Art. 28); consequently, to divide any quantity by a fraction, we invert the terms of the divisor, and multiply by the resulting fraction. Hence,

$$\frac{a}{b} \div \frac{c}{d} = \frac{a}{b} \times \frac{d}{c} = \frac{ad}{bc}.$$

Whence, the following rule for dividing one fraction by another:

### RULE.

I. *Reduce mixed quantities to fractional forms:*

II. *Invert the terms of the divisor, and multiply the dividend by the resulting fraction.*

NOTE.—The same remarks as were made on *factoring* and *reducing*, under the head of Multiplication, are applicable in Division.

### EXAMPLES.

1. Divide $a - \dfrac{b}{2c}$ by $\dfrac{f}{g}$.

$$a - \frac{b}{2c} = \frac{2ac - b}{2c}$$

Hence, $a - \dfrac{b}{2c} \div \dfrac{f}{g} = \dfrac{2ac - b}{2c} \times \dfrac{g}{f} = \dfrac{2acg - bg}{2cf}.$ *Ans.*

---

94. What is the rule for the division of fractions?

**2.** Divide $\dfrac{2(x+y)}{a}$ by $\dfrac{x^2-y^2}{a}$.

$$\dfrac{2(x+y)}{a} \times \dfrac{a}{x^2-y^2} = \dfrac{2(x+y)}{a} \times \dfrac{a}{(x+y)(x-y)}$$

$$= \dfrac{2}{x-y}. \quad Ans.$$

**3.** Let $\dfrac{7x}{5}$ be divided by $\dfrac{12}{13}$. $\qquad$ *Ans.* $\dfrac{91x}{60}$.

**4.** Let $\dfrac{4x^2}{7}$ be divided by $5x$. $\qquad$ *Ans.* $\dfrac{4x}{35}$.

**5.** Let $\dfrac{x+1}{6}$ be divided by $\dfrac{2x}{3}$. $\qquad$ *Ans.* $\dfrac{x+1}{4x}$.

**6.** Let $\dfrac{x}{x-1}$ be divided by $\dfrac{x}{2}$. $\qquad$ *Ans.* $\dfrac{2}{x-1}$.

**7.** Let $\dfrac{5x}{3}$ be divided by $\dfrac{2a}{3b}$. $\qquad$ *Ans.* $\dfrac{5bx}{2a}$.

**8.** Let $\dfrac{x-b}{8cd}$ be divided by $\dfrac{3cx}{4d}$. $\qquad$ *Ans.* $\dfrac{x-b}{6c^2x}$.

Divide the following fractions:

**9.** $\dfrac{4x^2-8x}{3}$ by $\dfrac{x^2-4}{3}$. $\qquad$ *Ans.* $\dfrac{4x}{x+2}$.

**10.** $\dfrac{x^4-b^4}{x^2-2bx+b^2}$ by $\dfrac{x^2+bx}{x-b}$. $\qquad$ *Ans.* $x+\dfrac{b^2}{x}$.

**11.** $2x(a+b)$ by $\dfrac{4x^2}{a+b}$. $\qquad$ *Ans.* $\dfrac{(a+b)^2}{2x}$.

**12.** $\dfrac{x^2-1}{y}$ by $\dfrac{(x+1)y^2}{x-1}$. $\qquad$ *Ans.* $\dfrac{(x-1)^2}{y^3}$.

**13.** $\dfrac{a^2-ax}{bc+bx}$ by $\dfrac{3(c-x)}{4(a+x)}$. $\qquad$ *Ans.* $\dfrac{4a(a^2-x^2)}{3b(c^2-x^2)}$.

14. $\dfrac{a-x}{1-x}$ by $\dfrac{1+x}{a+x}$.          Ans. $\dfrac{a^2-x^2}{1-x^2}$.

15. $x^2$ by $x - \dfrac{2xy}{x+y}$.          Ans. $x + \dfrac{2xy}{x-y}$.

16. $\dfrac{b-3a}{2ab}$ by $\dfrac{6a-2b}{b^2-2ab}$.          Ans. $\dfrac{2a-b}{4a}$.

17. $\dfrac{x^4-y^4}{x^2y}$ by $\dfrac{x}{y}+\dfrac{y}{x}$.          Ans. $\dfrac{x^2-y^2}{x}$.

18. $m^2+1+\dfrac{1}{m^2}$ by $m+\dfrac{1}{m}+1$.

          Ans. $m+\dfrac{1}{m}-1$.

19. $\left(x+\dfrac{y-x}{1+xy}\right)$ by $\left(1-x\dfrac{y-x}{1+xy}\right)$.      Ans. $y$.

20. $\left(\dfrac{x+2y}{x+y}+\dfrac{x}{y}\right)$ by $\left(\dfrac{x+2y}{y}-\dfrac{x}{x+y}\right)$.  Ans. 1

# CHAPTER V.

## EQUATIONS OF THE FIRST DEGREE.

**95.** An Equation is the expression of equality between two quantities. Thus,

$$x = b + c,$$

is an equation, expressing the fact that the quantity $x$, is equal to the sum of the quantities $b$ and $c$.

**96.** Every equation is composed of two parts, connected by the sign of equality. These parts are called *members:* the part on the left of the sign of equality, is called the *first member;* that on the right, the *second member.* Thus, in the equation,

$$x + a = b - c,$$

$x + a$ is the first member, and $b - c$, the second member.

**97.** An equation of the *first degree* is one which involves only the first power of the unknown quantity; thus,

$$6x + 3x - 5 = 13; \quad (1)$$
$$\text{and} \quad ax + bx + c = d; \quad (2)$$

are equations of the first degree.

---

95. What is an equation?

96. Of how many parts is every equation composed? How are the parts connected? What are the parts called? What is the part on the left called? The part on the right?

97. What is an equation of the first degree?

5\*

**98.** A NUMERICAL EQUATION is one in which the coefficients of the unknown quantity are denoted by numbers.

**99.** A LITERAL EQUATION is one in which the coefficients of the unknown quantity are denoted by letters.

Equation (1) is a numerical equation; Equation (2) is a literal equation.

EQUATIONS OF THE FIRST DEGREE CONTAINING BUT ONE UNKNOWN QUANTITY.

**100.** The TRANSFORMATION of an equation, is the operation of changing its form without destroying the equality of its members.

**101.** An AXIOM is a self-evident proposition.

**102.** The transformation of equations depends upon the following axioms:

1. *If equal quantities be added to both members of an equation, the equality will not be destroyed.*

2. *If equal quantities be subtracted from both members of an equation, the equality will not be destroyed.*

3. *If both members of an equation be multiplied by the same quantity, the equality will not be destroyed.*

4. *If both members of an equation be divided by the same quantity, the equality will not be destroyed.*

5. *Like powers of the two members of an equation are equal.*

6. *Like roots of the two members of an equation are equal.*

---

98. What is a numerical equation?
99. What is a literal equation?
100. What is the transformation of an equation?
101. What is an axiom?
102. Name the axioms on which the transformation of an equation depends.

**103.** Two principal transformations are employed in the solution of equations of the first degree: *Clearing of fractions*, and *Transposing*.

### CLEARING OF FRACTIONS.

1. Take the equation,

$$\frac{2x}{3} - \frac{3x}{4} + \frac{x}{6} = 11.$$

The least common multiple of the denominators is 12. If we multiply both members of the equation by 12, each term will reduce to an entire form, giving,

$$8x - 9x + 2x = 132.$$

Any equation may be reduced to entire terms in the same manner.

**104.** Hence for clearing of fractions, we have the following

### RULE.

I. *Find the least common multiple of the denominators:*

II. *Multiply both members of the equation by it, reducing the fractional to entire terms.*

Note.—1. The reduction will be effected, if we divide the least common multiple by each of the denominators, and then multiply the corresponding numerator, dropping the denominator.

2. The transformation may be effected by multiplying each numerator into the product of all the denominators except its own, omitting denominators.

---

103. How many transformations are employed in the solution of equations of the first degree? What are they?

104. Give the rule for clearing an equation of fractions? In what three ways may the reduction be effected?

**3.** The transformation may also be effected, by *multiplying both members of the equation by any multiple of the denominators.*

EXAMPLES.

Clear the following equations of fractions:

1. $\frac{x}{5} + \frac{x}{7} - 4 = 3.$    *Ans.* $7x + 5x - 140 = 105.$

2. $\frac{x}{6} + \frac{x}{9} - \frac{x}{27} = 8.$    *Ans.* $9x + 6x - 2x = 432.$

3. $\frac{x}{2} + \frac{x}{3} - \frac{x}{9} + \frac{x}{12} = 20.$

*Ans.* $18x + 12x - 4x + 3x = 720.$

4. $\frac{x}{5} + \frac{x}{7} - \frac{x}{2} = 4.$  *Ans.* $14x + 10x - 35x = 280.$

5. $\frac{x}{4} - \frac{x}{5} + \frac{x}{6} = 15.$  *Ans.* $15x - 12x + 10x = 900.$

6. $-\frac{x-4}{3} - \frac{x-2}{6} = \frac{5}{3}.$

*Ans.* $-2x + 8 - x + 2 = 10.$

7. $\frac{x}{3-x} + 4 = \frac{3}{5}.$  *Ans.* $5x + 60 - 20x = 9 - 3x.$

8. $\frac{x}{4} - \frac{x}{6} + \frac{x}{8} + \frac{x}{9} = 12.$

*Ans.* $18x - 12x + 9x + 8x = 864.$

9. $\frac{a}{b} - \frac{c}{d} + f = g.$    *Ans.* $ad - bc + bdf = bdg$

10. $\frac{ax}{b} - \frac{2c^2x}{ab} + 4a = \frac{4bc^2x}{a^3} - \frac{5a^3}{b^2} + \frac{2c^2}{a} - 3b.$

The least common multiple of the denominators is $a^3b^2$;

$a^4bx - 2a^2bc^2x + 4a^4b^2 = 4b^3c^2x - 5a^6 + 2a^2b^2c^2 - 3a^3b^3$

TRANSPOSING.

**105.** TRANSPOSITION is the operation of changing a term from one member to the other, without destroying the equality of the members.

1. Take, for example, the equation,

$$5x - 6 = 8 + 2x.$$

If, in the first place, we subtract $2x$ from both members the equality will not be destroyed, and we have,

$$5x - 6 - 2x = 8.$$

Whence we see, that the term $2x$, which was additive in the second member, becomes subtractive by passing into the first.

In the second place, if we add 6 to both members of the last equation, the equality will still exist, and we have,

$$5x - 6 - 2x + 6 = 8 + 6,$$

or, since $-6$ and $+6$ cancel each other, we have,

$$5x - 2x = 8 + 6.$$

Hence, the term which was subtractive in the first member, passes into the second member with the sign of addition.

**106.** Therefore, for the transposition of the terms, we have the following

### R U L E.

*Any term may be transposed from one member of an equation to the other, if the sign be changed.*

---

105. What is transposition?
106. What is the rule for the transposition of the terms of an equation?

Transpose the unknown terms to the first member, and the known terms to the second, in the following:

1. $3x + 6 - 5 = 2x - 7$.   *Ans.* $3x - 2x = -7 - 6 + 5$.
2. $ax + b = d - cx$.        *Ans.* $ax + cx = d - b$.
3. $4x - 3 = 2x + 5$.        *Ans.* $4x - 2x = 5 + 3$.
4. $9x + c = cx - d$.        *Ans.* $9x - cx = -d - c$.
5. $ax + f = dx + b$.        *Ans.* $ax - dx = b - f$.
6. $6x - c = -ax + b$.       *Ans.* $6x + ax = b + c$.

---

### SOLUTION   OF   EQUATIONS.

**107.** The SOLUTION of an equation is the operation of finding such a value for the unknown quantity, as will *satisfy* the equation; that is, such a value as, being substituted for the unknown quantity, will render the two members equal. This is called a ROOT of the equation.

A *Root* of an equation is said to be *verified*, when being substituted for the unknown quantity in the given equation, the two members are found equal to each other.

1. Take the equation,

$$\frac{3x}{2} - 4 = \frac{4(x - 2)}{8} + 3.$$

Clearing of fractions (Art. 104), and performing the operations indicated, we have,

$$12x - 32 = 4x - 8 + 24.$$

---

107. What is the solution of an equation? What is the found value of the unknown quantity called? When is a root of an equation said to be verified.

Transposing all the unknown terms to the first member, and the known terms to the second (Art. 106), we have,

$$12x - 4x = -8 + 24 + 32.$$

Reducing the terms in the two members,

$$8x = 48.$$

Dividing both members by the coefficient of $x$,

$$x = \frac{48}{8} = 6.$$

$$\frac{3 \times 6}{2} - 4 = \frac{4(6 - 2)}{8} + 3 \; ; \; \text{or,}$$

$$+ 9 - 4 = 2 + 3 = 5.$$

Hence, 6 satisfies the equation, and therefore, is a root.

**108.** By processes similar to the above, all equations of the first degree, containing but one unknown quantity, may be solved.

RULE.

I. *Clear the equation of fractions, and perform all the indicated operations:*

II. *Transpose all the unknown terms to the first member, and all the known terms to the second member:*

III. *Reduce all the terms in the first member to a single term, one factor of which will be the unknown quantity, and the other factor will be the algebraic sum of its coefficients:*

IV. *Divide both members by the coefficient of the unknown quantity: the second member will then be the value of the unknown quantity.*

---

108. Give the rule for solving equations of the first degree with one unknown quantity.

## EXAMPLES.

**1.** Solve the equation,

$$\frac{5x}{12} - \frac{4x}{3} - 13 = \frac{7}{8} - \frac{13x}{6}.$$

Clearing of fractions,

$$10x - 32x - 312 = 21 - 52x.$$

By transposing,

$$10x - 32x + 52x = 21 + 312.$$

By reducing,       $30x = 333;$

hence,          $x = \dfrac{333}{30} = \dfrac{111}{10} = 11.1;$

a result which may be verified by substituting it for $x$ in the given equation.

**2.** Solve the equation,

$$(3a - x)(a - b) + 2ax = 4b(x + a).$$

Performing the indicated operations, we have,

$$3a^2 - ax - 3ab + bx + 2ax = 4bx + 4ab.$$

By transposing,

$$- ax + bx + 2ax - 4bx = 4ab + 3ab - 3a^2.$$

By reducing,      $ax - 3bx = 7ab - 3a^2;$

Factoring,       $(a - 3b)x = 7ab - 3a^2.$

Dividing both members by the coefficient of $x$,

$$x = \frac{7ab - 3a^2}{a - 3b}.$$

**3.** Given $3x - 2 + 24 = 31$ to find $x$. *Ans.* $x = 3$.

**4.** Given $x + 18 = 3x - 5$ to find $x$. *Ans.* $x = 11\frac{1}{2}$.

**5.** Given $6 - 2x + 10 = 20 - 3x - 2$, to find $x$.

Ans. $x = 2$.

**6.** Given $x + \frac{1}{2}x + \frac{1}{3}x = 11$, to find $x$. Ans. $x = 6$.

**7.** Given $2x - \frac{1}{2}x + 1 = 5x - 2$, to find $x$.

Ans. $x = \frac{4}{5}$.

Solve the following equations:

**8.** $3ax + \frac{a}{2} - 3 = bx - a.$     Ans. $x = \dfrac{6 - 3a}{6a - 2b}.$

**9.** $\dfrac{x - 3}{2} + \dfrac{x}{3} = 20 - \dfrac{x - 19}{2}.$     Ans. $x = 23\frac{1}{4}.$

**10.** $\dfrac{x + 3}{2} + \dfrac{x}{3} = 4 - \dfrac{x - 5}{4}.$     Ans. $x = 3\frac{6}{13}.$

**11.** $\dfrac{x}{4} - \dfrac{3x}{2} + x = \dfrac{4x}{8} - 3.$     Ans. $x = 4.$

**12.** $\dfrac{3ax}{c} - \dfrac{2bx}{d} - 4 = f.$     Ans. $x = \dfrac{cdf + 4cd}{3ad - 2bc}.$

**13.** $\dfrac{x - a}{3} - \dfrac{2x - 3b}{5} - \dfrac{a - x}{2} = 10a + 11b.$

Ans. $x = 25a + 24b.$

**14.** $\dfrac{x}{12} - \dfrac{8 - x}{8} - \dfrac{5 + x}{4} + \dfrac{11}{4} = 0.$ Ans. $x = 12.$

**15.** $\dfrac{a + c}{a + x} + \dfrac{a - c}{a - x} = \dfrac{2b^2}{a^2 - x^2}$     Ans. $x = \dfrac{a^2 - b^2}{c}.$

**16.** $\dfrac{8ax - b}{7} - \dfrac{3b - c}{2} = 4 - b.$

Ans. $x = \dfrac{56 + 9b - 7c}{16a}.$

**17** $\dfrac{x}{5} - \dfrac{x - 2}{3} + \dfrac{x}{2} = \dfrac{13}{3}.$     Ans. $x = 10.$

18. $\dfrac{x}{a} - \dfrac{x}{b} + \dfrac{x}{c} - \dfrac{x}{d} = f.$

$$Ans.\ x = \frac{abcdf}{bcd - acd + abd - abc}.$$

NOTE.—What is the numerical value of $x$, when $a = 1$, $b = 2$, $c = 3$, $d = 4$, and $f = 6$?

19. $\dfrac{x}{7} - \dfrac{8x}{9} - \dfrac{x-3}{5} = -12\frac{22}{43} \cdots$　　　$Ans.\ x = 14.$

20. $x - \dfrac{3x-5}{13} + \dfrac{4x-2}{11} = x + 1.$　　$Ans.\ x = 6.$

21. $x + \dfrac{x}{4} + \dfrac{x}{5} - \dfrac{x}{6} = 2x - 43.$　　　$Ans.\ x = 60.$

22. $2x - \dfrac{4x-2}{5} = \dfrac{3x-1}{2}.$　　　　　$Ans.\ x = 3.$

23. $3x + \dfrac{bx-d}{3} = x + a.$　　$Ans.\ x = \dfrac{3a+d}{6+b}.$

24. $\dfrac{ax-b}{4} + \dfrac{a}{3} = \dfrac{bx}{2} - \dfrac{bx-a}{3}.$

$$Ans.\ x = \frac{3b}{3a - 2b}.$$

25. $\dfrac{4x}{5-x} - \dfrac{20-4x}{x} = \dfrac{15}{x}.$　　　$Ans.\ x = 3\dfrac{2}{11}.$

26. $\dfrac{2x+1}{29} - \dfrac{402-3x}{12} = 9 - \dfrac{471-6x}{2}.$

$$Ans.\ x = 72.$$

27. $\dfrac{(a+b)(x-b)}{a-b} - 3a = \dfrac{4ab-b^2}{a+b} - 2x + \dfrac{a^2-bx}{b}.$

$$Ans.\ x = \frac{a^4 + 3a^3b + 4a^2b^2 - 6ab^3 + 2b^4}{2b(2a^2 + ab - b^2)}.$$

## PROBLEMS.

**109.** A Problem is a question proposed, requiring a solution.

The Solution of a problem is the operation of finding a quantity, or quantities, that will satisfy the given conditions.

The solution of a problem consists of two parts:

I. The STATEMENT, *which consists in expressing, algebraically, the relation between the known and the required quantities.*

II. The SOLUTION, *which consists in finding the values of the unknown quantites, in terms of those which are known.*

The statement is made by representing the unknown quantities of the problem by some of the final letters of the alphabet, and then operating upon these so as to comply with the conditions of the problem. The method of stating problems is best learned by practical examples.

1. What number is that to which if 5 be added, the sum will be equal to 9 ?

Denote the number by $x$. Then, by the conditions,

$$x + 5 = 9.$$

This is the *statement* of the problem.

To find the value of $x$, transpose 5 to the second member; then,

$$x = 9 - 5 = 4.$$

This is the *solution* of the equation.

VERIFICATION.

$$x + 5 = 9.$$

---

109. What is a problem? What is the solution of a problem? Of how many parts does it consist? What are they? What is the statement? What is the solution?

2. Find a number such that the sum of one-half, one-third, and one-fourth of it, augmented by 45, shall be equal to 448

Let the required number be denoted by $x$.

Then, one-half of it will be denoted by $\dfrac{x}{2}$,

one-third    "    "    by $\dfrac{x}{3}$,

one-fourth    "    "    by $\dfrac{x}{4}$;

and, by the conditions,

$$\frac{x}{2} + \frac{x}{3} + \frac{x}{4} + 45 = 448.$$

This is the statement of the problem.

Clearing of fractions,

$$6x + 4x + 3x + 540 = 5376,$$

Transposing and collecting the unknown terms,

$$13x = 4836;$$

hence,    $$x = \frac{4836}{13} = 372.$$

VERIFICATION.

$$\frac{372}{2} + \frac{372}{3} + \frac{372}{4} + 45 = 186 + 124 + 93 + 45 = 448.$$

3. What number is that whose third part exceeds its fourth by 16?

Let the required number be denoted by $x$. Then,

$$\frac{1}{3}x = \text{the third part,}$$

$$\frac{1}{4}x = \text{the fourth part;}$$

and, by the conditions of the problem,

$$\frac{1}{3}x - \frac{1}{4}x = 16.$$

This is the statement. Clearing of fractions,

$$4x - 3x = 192,$$

and hence, $\qquad x = 192.$

VERIFICATION.

$$\frac{192}{3} - \frac{192}{4} = 64 - 48 = 16.$$

4. Divide \$1000 between $A$, $B$, and $C$, so that $A$ shall have \$72 more than $B$, and $C$ \$100 more than $A$.

Let $x$ denote the number of dollars which $B$ received.

Then, $\qquad x \qquad = B's$ number,

$\qquad\qquad x + 72 = A's$ number,

and, $\qquad x + 172 = C's$ number;

and their sum, $3x + 244 = 1000$, the number of dollars.

This is the statement. By transposing,

$$3x = 1000 - 244 = 756;$$

and, $\qquad x = \dfrac{756}{3} = 252 = B's$ share.

Hence, $x + 72 = 252 + 72 = 324 = A's$ share,

and, $\qquad x + 172 = 252 + 172 = 424 = C's$ share.

VERIFICATION.

$$252 + 324 + 424 = 1000.$$

5. Out of a cask of wine which had leaked away a third part, 21 gallons were afterwards drawn, and the cask being then guaged, appeared to be half full: how much did it hold?

Let    $x$ denote the number of gallons.

Then,    $\dfrac{x}{3}$ = the number that had leaked away.

and,    $\dfrac{x}{3} + 21$ = what had leaked and been drawn.

Hence, by the conditions, $\dfrac{x}{3} + 21 = \dfrac{x}{2}.$

This is the statement.    Clearing of fractions,

$$2x + 126 = 3x,$$

and,    $$- x = - 126 ;$$

and by changing the signs of both members, which does not destroy their equality (since it is equivalent to multiplying both members by $- 1$), we have,

$$x = 126.$$

### VERIFICATION.

$$\frac{126}{3} + 21 = 42 + 21 = 63 = \frac{126}{2}.$$

6. A fish was caught whose tail weighed 9 lbs., his head weighed as much as his tail and half his body, and his body weighed as much as his head and tail together: what was the weight of the fish?

Let    $2x$ = the weight of the body, in pounds.

Then,    $9 + x$ = weight of the head;

and since the body weighed as much as both head and tail,

$$2x = 9 + 9 + x,$$

which is the statement.    Then,

$$2x - x = 18, \text{ and } x = 18.$$

Hence, we have,

$$2x = 36lb. = \text{weight of the body,}$$
$$9 + x = 27lb. = \text{weight of the head,}$$
$$9lb. = \text{weight of the tail;}$$

hence,     $\overline{72lb.} = \text{weight of the fish.}$

7. The sum of two numbers is 67, and their differerce 19 what are the two numbers?

Let   $x$ denote the less number.

Then, $x + 19 =$ the greater; and, by the conditions,

$$2x + 19 = 67.$$

This is the statement. Transposing,

$$2x = 67 - 19 = 48;$$

hence,     $x = \dfrac{48}{2} = 24,$ and $x + 19 = 43.$

#### VERIFICATION.

$$43 + 24 = 67, \text{ and } 43 - 24 = 19.$$

#### ANOTHER SOLUTION.

Let   $x$   denote the greater number.

Then,   $x - 19$ will represent the less,

and,   $2x - 19 = 67;$ whence $2x = 67 + 19.$

Therefore,     $x = \dfrac{86}{2} = 43;$

and, consequently, $x - 19 = 43 - 19 = 24.$

#### GENERAL SOLUTION OF THIS PROBLEM.

The sum of two numbers is $s$, their difference is $d$: what are the two numbers?

Let      $x$      denote the *less* number.

Then,     $x + d$   will denote the greater,

and      $2x + d = s$, their sum. Whence,

$$x = \frac{s - d}{2} = \frac{s}{2} - \frac{d}{2};$$

and, consequently,

$$x + d = \frac{s}{2} - \frac{d}{2} + d = \frac{s}{2} + \frac{d}{2}.$$

As these two results are not dependent on particular values attributed to $s$ or $d$, it follows that:

1. *The greater of two numbers is equal to half their sum, plus half their difference:*

2. *The less is equal to half their sum, minus half their difference.*

Thus, if the sum of two numbers is 32, and their difference 16,

the greater is,    $\dfrac{32}{2} + \dfrac{16}{2} = 16 + 8 = 24$;   and

the less,      $\dfrac{32}{2} - \dfrac{16}{2} = 16 - 8 = 8.$

VERIFICATION.

$24 + 8 = 32$; and $24 - 8 = 16.$

8. A person engaged a workman for 48 days. For each day that he labored he received 24 cents, and for each day that he was idle, he paid 12 cents for his board. At the end of the 48 days, the account was settled, when the laborer received 504 cents. *Required, the number of working days, and the number of days he was idle.*

If the number of working days, and the number of idle days, were known, and the first multiplied by 24, and the

second by 12, the difference of these products would•be.
504. Let us indicate these operations by means of algebraic
signs.

Let $x$ denote the number of working days.

Then, $48 - x =$ the number of idle days,

$24 \times x =$ the amount earned,

and, $12(48 - x) =$ the amount paid for board.

Then, $24x - 12(48 - x) = 504$,

what was received, which is the statement.

Then, performing the operations indicated,

$$24x - 576 + 12x = 504,$$

or, $$36x = 504 + 576 = 1080,$$

and, $$x = \frac{1080}{36} = 30,$$ the number of working days;

whence, $48 - 30 = 18$, the number of idle days.

VERIFICATION.

Thirty days' labor, at 24 cents a day, amounts to . . . . . $\Big\}$ $30 \times 24 = 720$ cents.

And 18 days' board, at 12 cents a day, amounts to . . . . . $\Big\}$ $18 \times 12 = \underline{216}$ cents.

The difference is the amount received . . . . 504 cents.

GENERAL SOLUTION.

This problem may be made general, by denoting the whole number of working and idle days, by $n$;

The amount received for each day's work, by $a$;

The amount paid for board, for each idle day, by $b$;

And what was due the laborer, or the balance of the account, by $c$.

6

As before, let the number of working days be denoted by $x$.

The number of idle days will then be denoted by $n - x$.

Hence, what is earned will be expressed by $ax$, and the sum to be deducted, on account of board, by $b(n - x)$.

The statement of the problem, therefore, is,

$$ax - b(n - x) = c.$$

Performing indicated operations,

$$ax - bn + bx = c, \quad \text{or,} \quad (a + b)x = c + bn;$$

whence, $\qquad x = \dfrac{c + bn}{a + b} = $ number of working days;

and, $\qquad n - x = n - \dfrac{c + bn}{a + b} = \dfrac{an + bn - c - bn}{\cdot a + b},$

or, $\qquad n - x = \dfrac{an - c}{a + b} = $ number of idle days.

Let us suppose $n = 48$, $a = 24$, $b = 12$, and $c = 504$; these numbers will give for $x$ the same value as before found.

9. A person dying leaves half of his property to his wife, one-sixth to each of two daughters, one-twelfth to a servant, and the remaining $600 to the poor; what was the amount of the property?

Let $\qquad x$ denote the amount, in dollars,

Then, $\qquad \dfrac{x}{2} = \qquad$ what he left to his wife,

$\qquad\qquad \dfrac{x}{6} = \qquad$ what he left to one daughter,

and, $\qquad \dfrac{2x}{6} = \dfrac{x}{3}$ what he left to both daughters,

also, $\qquad \dfrac{x}{12} = \qquad$ what he left to his servant,

and, $\qquad$ $600 = \qquad$ what he left to the poor.

Then, by the conditions,

$$\frac{x}{2} + \frac{x}{3} + \frac{x}{12} + 600 = x,$$ the amount of the property,

which gives, $x = \$7200.$

10. $A$ and $B$ play together at cards. $A$ sits down with $84, and $B$ with $48. Each loses and wins in turn, when it appears that $A$ has five times as much as $B$. How much did $A$ win?

Let $x$ denote the number of dollars $A$ won.

Then, $A$ rose with $84 + x$ dollars,

and $B$ rose with $48 - x$ dollars.

But, by the conditions, we have,

$$84 + x = 5(48 - x),$$

hence, $84 + x = 240 - 5x;$

and, $6x = 156,$

consequently, $x = 26;$ or $A$ won $26.

VERIFICATION.

$$84 + 26 = 110; \quad 48 - 26 = 22;$$
$$110 = 5(22) = 110.$$

11. $A$ can do a piece of work alone in 10 days, $B$ in 13 days; in what time can they do it if they work together?

Denote the time by $x$, and the work to be done, by 1. Then, in

1 day, $A$ can do $\dfrac{1}{10}$ of the work, and

$B$ can do $\dfrac{1}{13}$ of the work; and in

$x$ days, $A$ can do $\dfrac{x}{10}$ of the work, and

$B$ can do $\dfrac{x}{13}$ of the work.

# 124 ELEMENTARY ALGEBRA.

Hence, by the conditions,

$$\frac{x}{10} + \frac{x}{13} = 1, \text{ which gives, } 13x + 10x = 130;$$

hence, $23x = 130$, $x = \dfrac{130}{23} = 5\frac{15}{23}$ days.

12. A fox, pursued by a hound, has a start of 60 of his own leaps. Three leaps of the hound are equivalent to 7 of the fox; but while the hound makes 6 leaps, the fox makes 9: how many leaps must the hound make to overtake the fox?

There is some difficulty in this problem, arising from the different units which enter into it.

Since 3 leaps of the hound are equal to 7 leaps of the fox, 1 leap of the hound is equal to $\dfrac{7}{3}$ fox leaps.

Since, while the hound makes 6 leaps, the fox makes 9, while the hound makes 1 leap, the fox will make $\dfrac{9}{6}$, or $\dfrac{3}{2}$ leaps.

Let $x$ denote the *number* of leaps which the hound makes before he overtakes the fox; and let 1 fox leap denote the *unit of distance.*

Since 1 leap of the hound is equal to $\dfrac{7}{3}$ of a fox leap, $x$ leaps will be equal to $\dfrac{7}{3}x$ fox leaps; and this will denote the distance passed over by the hound, in fox leaps.

Since, while the hound makes 1 leap, the fox makes $\dfrac{3}{2}$ leaps, while the hound makes $x$ leaps, the fox makes $\dfrac{3}{2}x$ leaps; and this added to 60, his distance ahead, will give $\dfrac{3}{2}x + 60$, for the whole distance passed over by the fox.

Hence, from the conditions,

$$\frac{7}{3}x = \frac{3}{2}x + 60; \text{ whence,}$$

$$14x = 9x + 360;$$

$$x = 72.$$

The hound, therefore, makes 72 leaps before overtaking the fox; in the same time, the fox makes $72 \times \frac{3}{2} = 108$ leaps.

VERIFICATION.

$$108 + 60 = 168, \text{ whole number of fox leaps,}$$

$$72 \times \frac{7}{3} = 168.$$

13. A father leaves his property, amounting to $2520, to four sons, $A$, $B$, $C$, and $D$. $C$ is to have $360, $B$ as much as $C$ and $D$ together, and $A$ twice as much as $B$, less $1000: how much do $A$, $B$, and $D$ receive?

*Ans.* $A$, $760; $B$, $880; $D$, $520.

14. An estate of $7500 is to be divided among a widow, two sons, and three daughters, so that each son shall receive twice as much as each daughter, and the widow herself $500 more than all the children: what was her share, and what the share of each child?

*Ans.* $\left\{ \begin{array}{l} \text{Widow's share, } 4000. \\ \text{Each son's,} \quad\quad 1000. \\ \text{Each daughter's,} \quad 500. \end{array} \right.$

15. A company of 180 persons consists of men, women, and children. The men are 8 more in number than the women, and the children 20 more than the men and women together: how many of each sort in the company?

*Ans.* 44 men, 36 women, 100 children.

16. A father divides $2000 among five sons, so that each elder should receive $40 more than his next younger brother : what is the share of the youngest?    *Ans.* $320.

17. A purse of $2850 is to be divided among three persons, *A*, *B*, and *C*. *A's* share is to be to *B's* as 6 to 11, and *C* is to have $300 more than *A* and *B* together : what is each one's share?    *A's*, $450 ; *B's*, $825 ; *C's*, $1575.

18. Two pedestrians start from the same point and travel in the same direction ; the first steps twice as far as the second, but the second makes 5 steps while the first makes but one. At the end of a certain time they are 300 feet apart. Now, allowing each of the longer paces to be 3 feet, how far will each have traveled?
                                        *Ans.* 1st, 200 feet ; 2d, 500.

19. Two carpenters, 24 journeymen, and 8 apprentices received at the end of a certain time $144. The carpenters received $1 per day, each journeyman, half a dollar, and each apprentice, 25 cents : how many days were they employed?    *Ans.* 9 days.

20. A capitalist receives a yearly income of $2940 ; four-fifths of his money bears an interest of 4 per cent., and the remainder of 5 per cent. : how much has he at interest?
                                        *Ans.* $70000.

21. A cistern containing 60 gallons of water has three unequal cocks for discharging it ; the largest will empty it in one hour, the second in two hours, and the third, in three : in what time will the cistern be emptied if they all run together?    *Ans.* $32\frac{8}{11}$ min.

22. In a certain orchard, one-half are apple trees, one-fourth peach trees, one-sixth plum trees ; there are also, 120 cherry trees, and 80 pear trees : how many trees in the orchard?    *Ans.* 2400.

23. A farmer being asked how many sheep he had,

answered, that he had them in five fields; in the 1st he had $\frac{1}{4}$, in the 2d, $\frac{1}{4}$, in the 3d, $\frac{1}{4}$, and in the 4th, $\frac{1}{12}$, and in the 5th, 450 : how many had he? *Ans.* 1200.

24. My horse and saddle together are worth $132, and the horse is worth ten times as much as the saddle: what is the value of the horse? *Ans.* $120.

25. The rent of an estate is this year 8 per cent. greater than it was last. This year it is $1890 : what was it last year? *Ans.* $1750.

26. What number is that, from which if 5 be subtracted, $\frac{2}{3}$ of the remainder will be 40? *Ans.* 65.

27. A post is $\frac{1}{4}$ in the mud, $\frac{1}{3}$ in the water, and 10 feet above the water: what is the whole length of the post? *Ans.* 24 feet.

28. After paying $\frac{1}{4}$ and $\frac{1}{6}$ of my money, I had 66 guineas left in my purse: how many guineas were in it at first? *Ans.* 120.

29. A person was desirous of giving 3 pence apiece to some beggars, but found he had not money enough in his pocket by 8 pence; he therefore gave them each 2 pence and had 3 pence remaining: required the number of beggars. *Ans.* 11.

30. A person, in play, lost $\frac{1}{4}$ of his money, and then won 3 shillings; after which he lost $\frac{1}{3}$ of what he then had; and this done, found that he had but 12 shillings remaining: what had he at first? *Ans.* 20s.

31. Two persons, *A* and *B*, lay out equal sums of money in trade; *A* gains $126, and *B* loses $87, and *A's* money is then double of *B's* : what did each lay out? *Ans.* $300.

32. A person goes to a tavern with a certain sum of money in his pocket, where he spends 2 shillings: he then borrows as much money as he had left, and going to another tavern, he there spends 2 shillings also; then borrowing

again as much money as was left, he went to a third tavern, where likewise he spent 2 shillings, and borrowed as much as he had left: and again spending 2 shillings at a fourth tavern, he then had nothing remaining. What had he at first?                          *Ans.* 3s. 9d.

33. A tailor cut 19 yards from each of three equal pieces of cloth, and 17 yards from another of the same length, and found that the four remnants were together equal to 142 yards. How many yards in each piece?      *Ans.* 54.

34. A fortress is garrisoned by 2600 men, consisting of infantry, artillery, and cavalry. Now, there are nine times as many infantry, and three times as many artillery soldiers as there are cavalry. How many are there of each corps?

       *Ans.* 200 cavalry; 600 artillery; 1800 infantry.

35. All the journeyings of an individual amounted to 2970 miles. Of these he traveled $3\frac{1}{2}$ times as many by water as on horseback, and $2\frac{1}{4}$ times as many on foot as by water. How many miles did he travel in each way?

         *Ans.* 240 miles; 840 m.; 1890 m.

36. A sum of money was divided between two persons, *A* and *B*. *A's* share was to *B's* in the proportion of 5 to 3, and exceeded five-ninths of the entire sum by 50. What was the share of each?     *Ans.* *A's* share, 450; *B's*, 270.

37. Divide a number $a$ into three such parts that the second shall be $n$ times the first, and the third $m$ times as great as the first.

$$\text{1st, } \frac{a}{1+m+n}; \quad \text{2d, } \frac{na}{1+m+n}; \quad \text{3d, } \frac{ma}{1+m+n}.$$

38. A father directs that $1170 shall be divided among his three sons, in proportion to their ages. The oldest is twice as old as the youngest, and the second is one-third older than the youngest. How much was each to receive?

       *Ans.* $270, youngest; $360, second; $540, oldest.

39. Three regiments are to furnish 594 men, and each to furnish in proportion to its strength. Now, the strength of the first is to the second as 3 to 5; and that of the second to the third as 8 to 7. How many must each furnish?

*Ans.* 1st, 144 men; 2d, 240; 3d, 210.

40. Five heirs, $A$, $B$, $C$, $D$, and $E$, are to divide an inheritance of $5600. $B$ is to receive twice as much as $A$, and $200 more; $C$ three times as much as $A$, less $400; $D$ the half of what $B$ and $C$ receive together, and 150 more; and $E$ the fourth part of what the four others get, plus $475. How much did each receive?

$A$'s, $500; $B$'s, 1200; $C$'s, 1100; $D$'s, 1300; $E$'s, 1500.

41. A person has four casks, the second of which being filled from the first, leaves the first four-sevenths full. The third being filled from the second, leaves it one-fourth full, and when the third is emptied into the fourth, it is found to fill only nine-sixteenths of it. But the first will fill the third and fourth, and leave 15 quarts remaining. How many gallons does each hold?

*Ans.* 1st, 35 gal.; 2d, 15 gal.; 3d, 11¼ gal.; 4th, 20 gal.

42. A courier having started from a place, is pursued by a second after the lapse of 10 days. The first travels 4 miles a day, the other 9. How many days before the second will overtake the first? *Ans.* 8.

43. A courier goes 31¼ miles every five hours, and is followed by another after he had been gone eight hours. The second travels 22½ miles every three hours. How many hours before he will overtake the first? *Ans.* 42.

44. Two places are eighty miles apart, and a person leaves one of them and travels towards the other at the rate of 3½ miles per hour. Eight hours after, a person departs from

6*

the second place, and travels at the rate of $5\frac{1}{6}$ miles per hour. How long before they will be together?

*Ans.* 6 hours.

EQUATIONS CONTAINING TWO UNKNOWN QUANTITIES.

**110.** If we have a single equation, as,

$$2x + 3y = 21,$$

containing two unknown quantities, $x$ and $y$, we may find the value of one of them in terms of the other, as,

$$x = \frac{21 - 3y}{2} \quad \cdots \cdots \quad (1.)$$

Now, if the value of $y$ is unknown, that of $x$ will also be unknown. Hence, from a *single* equation, containing two unknown quantities, the value of $x$ cannot be determined.

If we have a second equation, as,

$$5x + 4y = 35,$$

we may, as before, find the value of $x$ in terms of $y$, giving,

$$x = \frac{35 - 4y}{5} \quad \cdots \cdots \quad (2.)$$

Now, if the values of $x$ and $y$ are the same in Equations (1) and (2), the second members may be placed equal to each other, giving,

$$\frac{21 - 3y}{2} = \frac{35 - 4y}{5}, \quad \text{or} \quad 105 - 15y = 70 - 8y;$$

from which we find, $y = 5.$

---

110. In one equation containing two unknown quantities, can you find the value of either? If you have a second equation involving the same two unknown quantities, can you find their values? What are such equations called?

Subtituting this value for $y$ in Equations ($1$) or ($2$), we find $x = 3$. Such equations are called *Simultaneous equations.* Hence,

**111.** SIMULTANEOUS EQUATIONS are those in which the values of the unknown quantity are the same in both.

ELIMINATION.

**112.** ELIMINATION is the operation of combining two equations, containing two unknown quantities, and deducing therefrom a single equation, containing but one.

There are three principal methods of elimination :

1st. By addition or subtraction.
2d. By substitution.
3d. By comparison.

We shall consider these methods separately.

*Elimination by Addition or Subtraction.*

1. Take the two equations,

$$3x - 2y = 7,$$
$$8x + 2y = 48.$$

If we add these two equations, member to member, we obtain,

$$11x = 55;$$

which gives, by dividing by 11,

$$x = 5;$$

and substituting this value in either of the given equations, we find,

$$y = 4.$$

---

111. What are simultaneous equations?
112. What is elimination?   How many methods of elimination are there?   What are they?

2. Again, take the equations,

$$8x + 2y = 48,$$
$$3x + 2y = 23.$$

If we subtract the 2d equation from the 1st, we obtain,

$$5x = 25;$$

which gives, by dividing by 5,

$$x = 5;$$

and by substituting this value, we find,

$$y = 4.$$

3. Given the sum of two numbers equal to $s$, and their difference equal to $d$, to find the numbers.

Let $x =$ the greater, and $y$ the less number.

Then, by the conditions, . . . . . . $x + y = s.$
and, . . . . . . . . . . . . . $x - y = d.$
By adding (Art. 102, Ax. 1), . . . . . $2x = s + d.$
By subtracting (Art. 102, Ax. 2), . . . $2y = s - d.$

Each of these equations contains but one unknown quantity.

From the first, we obtain, . . . . . . $x = \dfrac{s + d}{2},$

and from the second, . . . . . . . . $y = \dfrac{s - d}{2}.$

These are the same values as were found in Prob. 7, page 120.

4. A person engaged a workman for 48 days. For each day that he labored he was to receive 24 cents, and for each day that he was idle he was to pay 12 cents for his board. At the end of the 48 days the account was settled, when the laborer received 504 cents. Required the number of working days, and the number of days he was idle.

Let $\qquad$ $x =$ the number of working days,

$\qquad\qquad$ $y =$ the number of idle days.

Then, $\qquad$ $24x =$ what he earned,

and, $\qquad$ $12y =$ what he paid for his board.

Then, by the conditions of the question, we have,

$$x + y = 48,$$

and, $\qquad\qquad 24x - 12y = 504.$

This is the statement of the problem.

It has already been shown (Art. 102, Ax. 3), that the two members of an equation may be multiplied by the same number, without destroying the equality. Let, then, the first equation be multiplied by 24, the coefficient of $x$ in the second; we shall then have,

$$24x + 24y = 1152$$
$$24x - 12y = 504$$

and by subtracting, $\qquad\qquad 36y = 648$

$$\therefore \; y = \frac{648}{36} = 18.$$

Substituting this value of $y$ in the equation,

$24x - 12y = 504,$ we have, $24x - 216 = 504;$

which gives,

$$24x = 504 + 216 = 720, \quad \text{and} \quad x = \frac{720}{24} = 30.$$

VERIFICATION.

$x + y = 48$ gives $\qquad\qquad 30 + 18 = 48,$

$24x - 12y = 504$ gives $\quad 24 \times 30 - 12 \times 18 = 504.$

**113.** In a similar manner, either unknown quantity may be eliminated from either equation; hence, the following

RULE.

I. *Prepare the equations so that the coefficients of the quantity to be eliminated shall be numerically equal:*

II. *If the signs are unlike, add the equations, member to member; if alike, subtract them, member from member.*

EXAMPLES.

Find the values of $x$ and $y$, by addition or subtraction, in the following simultaneous equations:

5. $\begin{cases} 3x - y = 3 \\ y + 2x = 7 \end{cases}$     *Ans.* $x = 2,\ y = 3.$

6. $\begin{cases} 4x - 7y = -22 \\ 5x + 2y = 37 \end{cases}$     *Ans.* $x = 5,\ y = 6.$

7. $\begin{cases} 2x + 6y = 42 \\ 8x - 6y = 3 \end{cases}$     *Ans.* $x = 4\frac{1}{2},\ y = 5\frac{1}{4}.$

8. $\begin{cases} 8x - 9y = 1 \\ 6x - 3y = 4x \end{cases}$     *Ans.* $x = \frac{1}{2},\ y = \frac{1}{3}.$

9. $\begin{cases} 14x - 15y = 12 \\ 7x + 8y = 37 \end{cases}$     *Ans.* $x = 3,\ y = 2.$

10. $\begin{cases} \frac{1}{2}x + \frac{1}{3}y = 6 \\ \frac{1}{3}x + \frac{1}{2}y = 6\frac{1}{2} \end{cases}$     *Ans.* $\begin{cases} x = 6,\ y = 9. \end{cases}$

11. $\begin{cases} \frac{1}{7}x + \frac{1}{8}y = 4 \\ x - y = -2 \end{cases}$     *Ans.* $\begin{cases} x = 14,\ y = 16. \end{cases}$

---

113. What is the rule for elimination by addition or subtraction?

12. Says *A* to *B*, you give me $40 of your money, and I shall then have five times as much as you will have left. Now they both had $120: how much had each?

*Ans.* Each had $60.

13. A father says to his son, "twenty years ago, my age was four times yours; now it is just double:" what were their ages?

*Ans.* { Father's, 60 years.<br>{ Son's, 30 years.

14. A father divided his property between his two sons. At the end of the first year the elder had spent one-quarter of his, and the younger had made $1000, and their property was then equal. After this the elder spent $500, and the younger made $2000, when it appeared that the younger had just double the elder: what had each from the father?

*Ans.* { Elder, $4000.<br>{ Younger, $2000.

15. If John give Charles 15 apples, they will have the same number; but if Charles give 15 to John, John will have 15 times as many, wanting 10, as Charles will have left. How many has each?

*Ans.* { John, 50.<br>{ Charles, 20.

16. Two clerks, *A* and *B*, have salaries which are together equal to $900. *A* spends $\frac{1}{10}$ per year of what he receives, and *B* adds as much to his as *A* spends. At the end of the year they have equal sums: what was the salary of each?

*Ans.* { *A's* = $500.<br>{ *B's* = $400.

### Elimination by Substitution.

**114.** Let us again take the equations,

$$5x + 7y = 43, \qquad (1.)$$
$$11x + 9y = 69. \qquad (2.)$$

---

Find the value of $x$ in the first equation, which gives,

$$x = \frac{43 - 7y}{5}.$$

*Substitute* this value of $x$ in the second equation, and we have,

$$11 \times \frac{43 - 7y}{5} + 9y = 69;$$

or,. $\quad\quad 473 - 77y + 45y = 345;$

or, $\quad\quad\quad\quad -32y = -128.$

Here, $x$ has been eliminated by *substitution*.

In a similar manner, we can eliminate any unknown quantity; hence, the

### RULE.

I. *Find from either equation the value of the unknown quantity to be eliminated:*

II. *Substitute this value for that quantity in the other equation.*

NOTE.—This method of elimination is used to great advantage when the coefficient of either of the unknown quantities is 1.

### EXAMPLES.

Find, by the last method, the values of $x$ and $y$ in the following equations:

1. $3x - y = 1$, and $3y - 2x = 4.$
$$\textit{Ans. } x = 1, \ y = 2.$$

2. $5y - 4x = -22$, and $3y + 4x = 38.$
$$\textit{Ans. } x = 8, \ y = 2.$$

3. $x + 8y = 18$, and $y - 3x = -29.$
$$\textit{Ans. } x = 10, \ y = 1.$$

4. $5x - y = 13$, and $8x + \frac{2}{9}y = 29$.

$\qquad\qquad$ *Ans.* $x = 3\frac{1}{2}$, $y = 4\frac{1}{2}$.

5. $10x - \frac{y}{5} = 69$, and $10y - \frac{x}{7} = 49$.

$\qquad\qquad$ *Ans.* $x = 7$, $y = 5$.

6. $x + \frac{1}{2}x - \frac{y}{5} = 10$, and $\frac{x}{8} + \frac{y}{10} = 2$.

$\qquad\qquad$ *Ans.* $x = 8$, $y = 10$.

7. $\frac{y}{7} - \frac{x}{3} + 5 = 2$, $x + \frac{y}{5} = 17\frac{4}{5}$.

$\qquad\qquad$ *Ans.* $x = 15$, $y = 14$.

8. $\frac{y}{2} + \frac{x}{3} + 3 = 6\frac{1}{6}$, and $\frac{y}{4} - \frac{x}{7} = \frac{1}{2}$.

$\qquad\qquad$ *Ans.* $x = 3\frac{1}{2}$, $y = 4$.

9. $\frac{y}{8} - \frac{x}{4} + 6 = 5$, and $\frac{x}{12} - \frac{y}{16} = 0$.

$\qquad\qquad$ *Ans.* $x = 12$, $y = 16$.

10. $\frac{y}{7} - \frac{3x}{2} - 1 = -9$, and $5x - \frac{7y}{49} = 29$.

$\qquad\qquad$ *Ans.* $x = 6$, $y = 7$.

11. Two misers, $A$ and $B$, sit down to count over their money. They both have $20000, and $B$ has three times as much as $A$ : how much has each ?

$\qquad\qquad$ *Ans.* $\begin{cases} A, \ \$5000. \\ B, \$15000. \end{cases}$

12. A person has two purses. If he puts $7 into the first, the whole is worth three times as much as the second purse : but if he puts $7 into the second, the whole is worth five times as much as the first : what is the value of each purse ?

$\qquad\qquad$ *Ans.* 1st, $2 ; 2d, $3.

13. Two numbers have the following properties: if the first be multiplied by 6, the product will be equal to the second multiplied by 5; and 1 subtracted from the first leaves the same remainder as 2 subtracted from the second: what are the numbers?                    *Ans.* 5 and 6.

14. Find two numbers with the following properties: the first increased by 2 is $3\frac{1}{4}$ times as great as the second; and the second increased by 4 gives a number equal to half the first: what are the numbers?            *Ans.* 24 and 8.

15. A father says to his son, "twelve years ago, I was twice as old as you are now: four times your age at that time, plus twelve years, will express my age twelve years hence:" what were their ages?

$$Ans. \begin{cases} \text{Father, 72 years.} \\ \text{Son,} \quad 30 \quad \text{``} \end{cases}$$

## *Elimination by Comparison.*

**115.** Take the same equations,

$$5x + 7y = 43$$
$$11x + 9y = 69.$$

Finding the value of $x$ from the first equation, we have,

$$x = \frac{43 - 7y}{5};$$

and finding the value of $x$ from the second, we obtain,

$$x = \frac{69 - 9y}{11}.$$

---

115. Give the rule for elimination by comparison.

Let these two values of $x$ be placed equal to each other, and we have,

$$\frac{43 - 7y}{5} = \frac{69 - 9y}{11}.$$

Or, $\qquad 473 - 77y = 345 - 45y ;$

or, $\qquad - 32y = - 128.$

Hence, $\qquad y = 4.$

And, $\qquad x = \dfrac{69 - 36}{11} = 3.$

This method of elimination is called the method by *comparison*, for which we have the following

### RULE.

I. *Find, from each equation, the value of the same unknown quantity to be eliminated:*

II. *Place these values equal to each other.*

### EXAMPLES.

Find, by the last rule, the values of $x$ and $y$, from the following equations,

1. $3x + \dfrac{y}{5} + 6 = 42,$ and $y - \dfrac{x}{22} = 14\frac{1}{2}.$

$\qquad\qquad$ *Ans.* $x = 11, \; y = 15.$

2. $\dfrac{y}{4} - \dfrac{x}{7} + 5 = 6,$ and $\dfrac{y}{5} + 4 = \dfrac{x}{14} + 6.$

$\qquad\qquad$ *Ans.* $x = 28, \; y = 20.$

3. $\dfrac{y}{10} - \dfrac{x}{4} + \dfrac{22}{8} = 1,$ and $3y - x = 6.$

$\qquad\qquad$ *Ans.* $x = 9, \; y = 5.$

4. $y - 3 = \dfrac{1}{2}x + 5,$ and $\dfrac{x + y}{2} = y - 3\frac{1}{2}.$

$\qquad\qquad$ *Ans.* $x = 2, \; y = 9.$

**5.** $\frac{y-x}{3} + \frac{x}{2} = y - 2$, and $\frac{x}{8} + \frac{y}{7} = x - 13$.

$\qquad\qquad\qquad\qquad$ *Ans.* $x = 16, \; y = 7$.

**6.** $\frac{y+x}{2} + \frac{y-x}{2} = x - \frac{2y}{3}$, and $x + y = 16$.

$\qquad\qquad\qquad\qquad$ *Ans.* $x = 10, \; y = 6$.

**7.** $\frac{2x-3y}{5} = x - 2\frac{2}{5}$, $\; x - \frac{y-1}{2} = 0$.

$\qquad\qquad\qquad\qquad$ *Ans.* $x = 1, \; y = 2$.

**8.** $2y + 3x = y + 43$, $\; y - \frac{x-4}{3} = y - \frac{x}{5}$.

$\qquad\qquad\qquad\qquad$ *Ans.* $x = 10, \; y = 13$.

**9.** $4y - \frac{x-y}{2} = x + 18$, and $27 - y = x + y + 4$.

$\qquad\qquad\qquad\qquad$ *Ans.* $x = 9, \; y = 7$.

**10.** $1 - \frac{y-x}{6} + 4 = y - 16\frac{2}{3}$, $\; \frac{y}{5} - 2 = \frac{x}{5}$.

$\qquad\qquad\qquad\qquad$ *Ans.* $x = 10, \; y = 20$.

**116.** Having explained the principal methods of elimination, we shall add a few examples which may be solved by any one of them; and often indeed, it may be advantageous to employ them all, even in the same example.

### GENERAL EXAMPLES.

Find the values of $x$ and $y$ in the following simultaneous equations:

**1.** $2x + 3y = 16$, and $3x - 2y = 11$.

$\qquad\qquad\qquad\qquad$ *Ans.* $x = 5, \; y = 2$.

**2.** $\dfrac{2x}{5} + \dfrac{3y}{4} = \dfrac{9}{20}$, and $\dfrac{3x}{4} + \dfrac{2y}{5} = \dfrac{61}{120}$.

$Ans.\ x = \dfrac{1}{2},\ y = \dfrac{1}{3}$.

**3.** $\dfrac{x}{7} + 7y = 99$, and $\dfrac{y}{7} + 7x = 51$.

$Ans.\ x = 7,\ y = 14$.

**4.** $\dfrac{x}{2} - 12 = \dfrac{y}{4} + 8$, $\dfrac{x+y}{5} + \dfrac{x}{3} - 8 = \dfrac{2y-x}{4} + 27$.

$Ans.\ x = 60,\ y = 40$.

**5.** $\left\{ \begin{aligned} x - \tfrac{1}{2}y + \dfrac{4x}{5} &= 6\tfrac{1}{2} \\[2mm] \dfrac{x - y}{2} + 7x &= 41 \end{aligned} \right\}$ $\qquad Ans.\ \left\{ \begin{aligned} x &= 6. \\[2mm] y &= 8. \end{aligned} \right.$

**6.** $\left\{ \begin{aligned} \dfrac{x - y}{4} + \dfrac{x + y}{5} &= 2\tfrac{1}{10} \\[2mm] \tfrac{1}{2}x - y + 4\tfrac{1}{4}y &= 12\tfrac{1}{4} \end{aligned} \right\}$ $\qquad Ans.\ \left\{ \begin{aligned} x &= 5. \\[2mm] y &= 3. \end{aligned} \right.$

**7.** $\left\{ \begin{aligned} \dfrac{3y - x}{6} + \dfrac{2x - y}{4} &= 5 \\[2mm] 6x - y + \dfrac{8 - 2x}{4} &= 43\tfrac{1}{2} \end{aligned} \right\}$ $\qquad Ans.\ \left\{ \begin{aligned} x &= 9. \\[2mm] y &= 8. \end{aligned} \right.$

**8.** $\left\{ \begin{aligned} \dfrac{3x - 8}{4} + \dfrac{y - 6}{5} + y &= 18\tfrac{7}{10} \\[2mm] 8x - 3 - \dfrac{6 - y}{3} &= 79 \end{aligned} \right\}$ $\qquad Ans.\ \left\{ \begin{aligned} x &= 10. \\[2mm] y &= 12. \end{aligned} \right.$

**9.** $\left\{ \begin{aligned} \dfrac{4x - 4}{3} - \dfrac{y - 5}{4} + 6 &= 12\tfrac{2}{3} \\[2mm] \tfrac{1}{4}x - \tfrac{1}{3}y + \dfrac{y - 4}{3} &= \tfrac{4}{3} \end{aligned} \right\}$ $\qquad Ans.\ \left\{ \begin{aligned} x &= 6. \\[2mm] y &= 5. \end{aligned} \right.$

10. $\begin{cases} ax - by = c \\ a - y + x = d \end{cases}$    *Ans.* $\begin{cases} x = \dfrac{c + ab - bd}{a - b}. \\ y = \dfrac{a^2 + c - ad}{a - b}. \end{cases}$

11. $\begin{cases} 13x + 7y - 341 = 7\frac{1}{2}y + 43\frac{1}{2}x \\ 2x + \frac{1}{2}y = 1 \end{cases}$   *Ans.* $\begin{cases} x = -12. \\ y = 50. \end{cases}$

12. $\begin{cases} (x+5)(y+7) = (x+1)(y-9)+112 \\ 2x+10 = 3y+1 \end{cases}$   *Ans.* $\begin{cases} x = 3. \\ y = 5. \end{cases}$

13. $\begin{cases} ax = by \\ x + y = c \end{cases}$    *Ans.* $\begin{cases} x = \dfrac{bc}{a + b}. \\ y = \dfrac{ac}{a + b}. \end{cases}$

14. $\begin{cases} ax + by = c \\ fx + gy = h \end{cases}$    *Ans.* $\begin{cases} x = \dfrac{cg - bh}{ag - bf} \\ y = \dfrac{ah - cf}{ag - bf}. \end{cases}$

15. $\begin{cases} \dfrac{a}{b + y} = \dfrac{b}{3a + x} \\ ax + 2by = d \end{cases}$    *Ans.* $\begin{cases} x = \dfrac{2b^2 - 6a^2 + d}{3a}. \\ y = \dfrac{3a^2 - b^2 + d}{3b}. \end{cases}$

16. $\begin{cases} bcx = cy - 2b \\ b^2y + \dfrac{a(c^3 - b^3)}{bc} = \dfrac{2b^3}{c} + c^3x \end{cases}$    *Ans.* $\begin{cases} x = \dfrac{a}{bc}. \\ y = \dfrac{a + 2b}{c}. \end{cases}$

17. $\begin{cases} 3x + 5y = \dfrac{(8b - 2f)bf}{b^2 - f^2} \\ y - x = \dfrac{-2bf^2}{b^2 - f^2} \end{cases}$    *Ans.* $\begin{cases} x = \dfrac{bf}{b - f}. \\ y = \dfrac{bf}{b + f}. \end{cases}$

## PROBLEMS.

1. What fraction is that, to the numerator of which if 1 be added, the value will be $\frac{1}{3}$, but if 1 be added to its denominator, the value will be $\frac{1}{4}$?

Let the fraction be denoted by $\frac{x}{y}$.

Then, by the conditions,

$$\frac{x+1}{y} = \frac{1}{3}, \text{ and, } \frac{x}{y+1} = \frac{1}{4}.$$

whence, $3x + 3 = y$, and $4x = y + 1$.

Therefore, by subtracting,

$$x - 3 = 1, \text{ and } x = 4.$$

Hence, $12 + 3 = y$;

$$\therefore \ y = 15.$$

2. A market-woman bought a certain number of eggs at 2 for a penny, and as many others at 3 for a penny; and having sold them all together, at the rate of 5 for 2$d$, found that she had lost 4$d$: how many of both kinds did she buy?

Let $\quad\quad 2x \quad$ denote the whole number of eggs.

Then, $\quad\quad x = \quad$ the number of eggs of each sort.

Then will, $\quad \frac{1}{2}x = \quad$ the cost of the first sort,

and, $\quad\quad \frac{1}{3}x = \quad$ the cost of the second sort.

But, by the conditions of the question,

$$5 : 2x :: 2 : \frac{4x}{5};$$

hence, $\frac{4x}{5}$ will denote the amount for which the eggs were sold.

But, by the conditions,

$$\frac{1}{2}x + \frac{1}{3}x - \frac{4x}{5} = 4;$$

therefore,          $15x + 10x - 24x = 120;$

$\therefore \ x = 120;$ the number of eggs of each sort.

3. A person possessed a capital of 30,000 dollars, for which he received a certain interest; but he owed the sum of 20,000 dollars, for which he paid a certain annual interest. The interest that he received exceeded that which he paid by 800 dollars. Another person possessed 35,000 dollars, for which he received interest at the second of the above rates; but he owed 24,000 dollars, for which he paid interest at the first of the above rates. The interest that he received, annually, exceeded that which he paid, by 310 dollars. Required the two rates of interest.

Let $x$ denote the number of units in the first rate of interest, and $y$ the unit in the second rate. Then each may be regarded as denoting the interest on $100 for 1 year.

To obtain the interest of $30,000 at the first rate, denoted by $x$, we form the proportion,

$$100 : 30,000 :: x : \frac{30,000x}{100}, \text{ or } 300x.$$

And for the interest of $20,000, the rate being $y$,

$$100 : 20,000 :: y : \frac{20,000y}{100}, \text{ or } 200y.$$

But, by the conditions, the difference between these two amounts is equal to 800 dollars.

We have, then, for the first equation of the problem,

$$300x - 200y = 800.$$

By expressing, algebraically, the second condition of the problem, we obtain a second equation,

$$350y - 240x = 310.$$

Both members of the first equation being divisible by 100, and those of the second by 10, we have,

$$3x - 2y = 8, \qquad 35y - 24x = 31.$$

To eliminate $x$, multiply the first equation by 8, and then add the result to the second; there results,

$$19y = 95, \quad \text{whence}, \quad y = 5.$$

Substituting for $y$, in the first equation, this value, and that equation becomes,

$$3x - 10 = 8, \quad \text{whence}, \quad x = 6.$$

Therefore, the first rate is 6 per cent, and the second 5.

VERIFICATION.

$30,000,    at 6 per cent, gives    $30,000 \times .06 = \$1800.$

$20,000,      5    "      "    $20,000 \times .05 = \$1000.$

And we have,    $1800 - 1000 = 800.$

The second condition can be verified in the same manner.

4. What two numbers are those, whose difference is 7, and sum 33 ?        *Ans.* 13 and 20.

5. Divide the number 75 into two such parts, that three times the greater may exceed seven times the less by 15.
       *Ans.* 54 and 21.

6. In a mixture of wine and cider, $\frac{1}{2}$ of the whole plus 25 gallons was wine, and $\frac{1}{4}$ part minus 5 gallons was cider: how many gallons were there of each ?
       *Ans.* 85 of wine, and 35 of cider.

7

7. A bill of £120 was paid in guineas and moidores, and the number of pieces used, of both sorts, was just 100. If the guinea be estimated at 21s, and the moidore at 27s, how many pieces were there of each sort ?      *Ans.* 50.

8. Two travelers set out at the same time from London and York, whose distance apart is 150 miles. One of them travels 8 miles a day, and the other 7 : in what time will they meet ?      *Ans.* In 10 days.

9. At a certain election, 375 persons voted for two candidates, and the candidate chosen had a majority of 91 : how many voted for each ?

*Ans.* 233 for one, and 142 for the other.

10. A person has two horses, and a saddle worth £50. Now, if the saddle be put on the back of the first horse, it makes their joint value double that of the second horse; but if it be put on the back of the second, it makes their joint value triple that of the first : what is the value of each horse?      *Ans.* One £30, and the other £40.

11. The hour and minute hands of a clock are exactly together at 12 o'clock : when will they be again together?

*Ans.* 1h. $5\frac{5}{11}$m.

12. A man and his wife usually drank out a cask of beer in 12 days; but when the man was from home, it lasted the woman 30 days : how many days would the man alone be in drinking it ?      *Ans.* 20 days.

13. If 32 pounds of sea-water contain 1 pound of salt, how much fresh water must be added to these 32 pounds, in order that the quantity of salt contained in 32 pounds of the new mixture shall be reduced to 2 ounces, or $\frac{1}{8}$ of a pound ?

*Ans.* 224 lbs.

14. A person who possessed 100,000 dollars, placed the greater part of it out at 5 per cent interest, and the other

at 4 per cent. The interest which he received for the whole, amounted to 4640 dollars. Required the two parts.

*Ans.* $64,000 and $36,000.

15. At the close of an election, the successful candidate had a majority of 1500 votes. Had a fourth of the votes of the unsuccessful candidate been also given to him, he would have received three times as many as his competitor, wanting three thousand five hundred : how many votes did each receive? *Ans.* { 1st, 6500. 2d, 5000.

16. A gentleman bought a gold and a silver watch, and a chain worth $25. When he put the chain on the gold watch, it and the chain became worth three and a half times more than the silver watch ; but when he put the chain on the silver watch, they became worth one-half the gold watch and 15 dollars over : what was the value of each watch? *Ans.* { Gold watch, $80. Silver " $30.

17. There is a certain number expressed by two figures, which figures are called digits. The sum of the digits is 11, and if 13 be added to the first digit the sum will be three times the second: what is the number? *Ans.* 56.

18. From a company of ladies and gentlemen 15 ladies retire; there are then left two gentlemen to each lady. After which 45 gentlemen depart, when there are left 5 ladies to each gentleman : how many were there of each at first? *Ans.* { 50 gentlemen. 40 ladies.

19. A person wishes to dispose of his horse by lottery. If he sells the tickets at $2 each, he will lose $30 on his horse ; but if he sells them at $3 each, he will receive $30

more than his horse cost him.  What is the value of the horse, and number of tickets?

$$Ans. \begin{cases} \text{Horse,} & \$150. \\ \text{No. of tickets, 60.} \end{cases}$$

20.  A person purchases a lot of wheat at $1, and a lot of rye at 75 cents per bushel; the whole costing him $117.50. He then sells $\frac{1}{4}$ of his wheat and $\frac{1}{3}$ of his rye at the same rate, and realizes $27.50.  How much did he buy of each?

$$Ans. \begin{cases} \text{80 bush. of wheat.} \\ \text{50 bush. of rye.} \end{cases}$$

21.  There are 52 pieces of money in each of two bags.  *A* takes from one, and *B* from the other.  *A* takes twice as much as *B* left, and *B* takes 7 times as much as *A* left. How much did each take?

$$Ans. \begin{cases} A, \text{ 48 pieces.} \\ B, \text{ 28 pieces.} \end{cases}$$

22.  Two persons, *A* and *B*, purchase a house together, worth $1200.  Says *A* to *B*, give me two-thirds of your money and I can purchase it alone; but, says *B* to *A*, if you will give me three-fourths of your money I shall be able to purchase it alone.  How much had each?

*Ans.* *A*, $800; *B*, $600.

23.  A grocer finds that if he mixes sherry and brandy in the proportion of 2 to 1, the mixture will be worth 78s. per dozen; but if he mixes them in the proportion of 7 to 2, he can get 79s. a dozen.  What is the price of each liquor per dozen?          *Ans.* Sherry, 81s.; brandy, 72s.

*Equations containing three or more unknown quantities.*

**117.**  Let us now consider equations involving three or more unknown quantities.

Take the group of simultaneous equations,

---

117. Give the rule for solving any group of simultaneous equations?

$$5x - 6y + 4z = 15, \quad . \quad . \quad (1.)$$
$$7x + 4y - 3z = 19, \quad . \quad . \quad (2.)$$
$$2x + y + 6z = 46. \quad . \quad . \quad . \quad (3.)$$

To eliminate $z$ by means of the first two equations, multiply the first by 3, and the second by 4; then, since the coefficients of $z$ have contrary signs, add the two results together. This gives a new equation:

$$43x - 2y = 121 \quad . \quad . \quad . \quad . \quad . \quad (4.)$$

Multiplying the second equation by 2 (a factor of the coefficient of $z$ in the third equation), and adding the result to the third equation, we have,

$$16x + 9y = 84 \quad . \quad . \quad . \quad . \quad . \quad (5.)$$

The question is then reduced to finding the values of $x$ and $y$, which will satisfy the new Equations (4) and (5).

Now, if the first be multiplied by 9, the second by 2, and the results added together, we find,

$$419x = 1257; \text{ whence, } x = 3.$$

We might, by means of Equations (4) and (5) determine $y$ in the same way that we have determined $x$; but the value of $y$ may be determined more simply, by substituting the value of $x$ in Equation (5); thus,

$$48 + 9y = 84. \qquad \therefore \quad y = \frac{84 - 48}{9} = 4.$$

In the same manner, the first of the three given equations becomes, by substituting the values of $x$ and $y$,

$$15 - 24 + 4z = 15. \qquad \therefore \quad z = \frac{24}{4} = 6.$$

In the same way, any group of simultaneous equations may be solved. Hence, the

## RULE.

I. *Combine one equation of the group with each of the others, by eliminating one unknown quantity; there will result a new group containing one equation less than the original group:*

II. *Combine one equation of this new group with each of the others, by eliminating a second unknown quantity; there will result a new group containing two equations less than the original group:*

III. *Continue the operation until a single equation is found, containing but one unknown quantity:*

IV. *Find the value of this unknown quantity by the preceding rules; substitute this in one of the group of two equations, and find the value of a second unknown quantity; substitute these in either of the group of three, finding a third unknown quantity; and so on, till the values of all are found.*

NOTES.—1. In order that the value of the unknown quantities may be determined, there must be just as many independent equations of condition as there are unknown quantities. If there are fewer equations than unknown quantities, the resulting equation will contain at least two unknown quantities, and hence, their values cannot be found (Art. 110). If there are more equations than unknown quantities, the conditions may be contradictory, and the equations impossible.

2. It often happens that each of the proposed equations does not contain all the unknown quantities. In this case, with a little address, the elimination is very quickly performed.

Take the four equations involving four unknown quantities:

$$2x - 3y + 2z = 13. \quad (1.) \qquad 4y + 2z = 14. \quad (3.)$$
$$4u - 2x = 30. \quad (2.) \qquad 5y + 3u = 32. \quad (4.)$$

By inspecting these equations, we see that the elimination of $z$ in the two Equations, ( 1 ) and ( 3 ), will give an equation involving $x$ and $y$; and if we eliminate $u$ in Equations ( 2 ) and ( 4 ), we shall obtain a second equation, involving $x$ and $y$. These last two unknown quantities may therefore be easily determined. In the first place, the elimination of $z$ from ( 1 ) and ( 3 ) gives,

$$7y - 2x = 1;$$

That of $u$ from ( 2 ) and ( 4 ) gives,

$$20y + 6x = 38.$$

Multiplying the first of these equations by 3, and adding,

$$41y = 41;$$

Whence, $\qquad y = 1.$

Substituting this value in $7y - 2x = 1$, we find,

$$x = 3.$$

Substituting for $x$ its value in Equation ( 2 ), it becomes

$$4u - 6 = 30.$$

Whence, $\qquad u = 9.$

And substituting for $y$ its value in Equation ( 3 ), there results,

$$z = 5.$$

### EXAMPLES.

**1.** Given $\begin{cases} x + y + z = 29 \\ x + 2y + 3z = 62 \\ \frac{1}{2}x + \frac{1}{3}y + \frac{1}{4}z = 10 \end{cases}$ to find $x$, $y$, and $z$.

$\qquad\qquad$ Ans. $x = 8,\ y = 9,\ z = 12.$

**2.** Given $\left\{\begin{array}{l} 2x + 4y - 3z = 22 \\ 4x - 2y + 5z = 18 \\ 6x + 7y - z = 63 \end{array}\right\}$ to find $x$, $y$, and $z$.

$$Ans.\ x = 3,\ y = 7,\ z = 4.$$

**3.** Given $\left\{\begin{array}{l} x + \dfrac{1}{2}y + \dfrac{1}{3}z = 32 \\ \dfrac{1}{3}x + \dfrac{1}{4}y + \dfrac{1}{5}z = 15 \\ \dfrac{1}{4}x + \dfrac{1}{5}y + \dfrac{1}{6}z = 12 \end{array}\right\}$ to find $x$, $y$, and $z$.

$$Ans.\ x = 12,\ y = 20,\ z = 30.$$

**4.** Given $\left\{\begin{array}{l} x + y + z = 29\frac{1}{4} \\ x + y - z = 18\frac{1}{4} \\ x - y + z = 13\frac{3}{4} \end{array}\right\}$ to find $x$, $y$, and $z$.

$$Ans.\ x = 16,\ y = 7\tfrac{3}{4},\ z = 5\tfrac{1}{2}.$$

**5.** Given $\left\{\begin{array}{l} 3x + 5y = 161 \\ 7x + 2z = 209 \\ 2y + z = 89 \end{array}\right\}$ to find $x$, $y$, and $z$.

$$Ans.\ x = 17,\ y = 22,\ z = 45.$$

**6.** Given $\left\{\begin{array}{l} \dfrac{1}{x} + \dfrac{1}{y} = a \\ \dfrac{1}{x} + \dfrac{1}{z} = b \\ \dfrac{1}{y} + \dfrac{1}{z} = c \end{array}\right\}$ to find $x$, $y$, and $z$.

$$x = \frac{2}{a + b - c},\quad y = \frac{2}{a + c - b},\quad z = \frac{2}{b + c - a}.$$

Note.—In this example we should not proceed to clear the equation of fractions; but subtract immediately the second equation from the first, and then add the third: we thus find the value of $y$.

## PROBLEMS.

1. Divide the number 90 into four such parts, that the first increased by 2, the second diminished by 2, the third multiplied by 2, and the fourth divided by 2, shall be equal each to each.

This problem may be easily solved by introducing a new unknown quantity.

Let $x$, $y$, $z$, and $u$, denote the required parts, and designate by $m$ the several equal quantities which arise from the conditions. We shall then have,

$$x + 2 = m, \quad y - 2 = m, \quad 2z = m, \quad \frac{u}{2} = m.$$

From which we find,

$$x = m - 2, \quad y = m + 2, \quad z = \frac{m}{2}, \quad u = 2m.$$

And, by adding the equations,

$$x + y + z + u = m + m + \frac{m}{2} + 2m = 4\tfrac{1}{2}m.$$

And since, by the conditions of the problem, the first member is equal to 90, we have,

$$4\tfrac{1}{2}m = 90, \quad \text{or} \quad \tfrac{9}{2}m = 90;$$

hence, $\qquad\qquad m = 20.$

Having the value of $m$, we easily find the other values; viz.:

$$x = 18, \quad y = 22, \quad z = 10, \quad u = 40.$$

2. There are three ingots, composed of different metals mixed together. A pound of the first contains 7 ounces of silver, 3 ounces of copper, and 6 of pewter. A pound of the second contains 12 ounces of silver, 3 ounces of copper, and 1 of pewter. A pound of the third contains 4 ounces of silver, 7 ounces of copper, and 5 of pewter. It is required

7*

to find how much it will take of each of the three ingots to form a fourth, which shall contain in a pound, 8 ounces of silver, $3\frac{3}{4}$ of copper, and $4\frac{1}{4}$ of pewter.

Let $x$, $y$, and $z$, denote the number of ounces which it is necessary to take from the three ingots respectively, in order to form a pound of the required ingot. Since there are 7 ounces of silver in a pound, or 16 ounces, of the first ingot, it follows that one ounce of it contains $\frac{7}{16}$ of an ounce of silver, and, consequently, in a number of ounces denoted by $x$, there is $\dfrac{7x}{16}$ ounces of silver. In the same manner, we find that, $\dfrac{12y}{16}$, and $\dfrac{4z}{16}$, denote the number of ounces of silver taken from the second and third; but, from the enunciation, one pound of the fourth ingot contains 8 ounces of silver. We have, then, for the first equation,

$$\frac{7x}{16} + \frac{12y}{16} + \frac{4z}{16} = 8;$$

or, clearing fractions,

$$7x + 12y + 4z = 128.$$

As respects the copper, we should find,

$$3x + 3y + 7z = 60;$$

and with reference to the pewter,

$$6x + y + 5z = 68.$$

As the coefficients of $y$ in these three equations are the most simple, it is convenient to eliminate this unknown quantity first.

Multiplying the second equation by 4, and subtracting the first from it, member from member, we have,

$$5x + 24z = 112.$$

Multiplying the third equation by 3, and subtracting the second from the resulting equation, we have,

$$15x + 8z = 144.$$

Multiplying this last equation by 3, and subtracting the preceding one, we obtain,

$$40x = 320;$$

whence, $\qquad x = 8.$

Substitute this value for $x$ in the equation,

$$15x + 8z = 144;$$

it becomes, $\qquad 120 + 8z = 144,$

whence, $\qquad z = 3.$

Lastly, the two values, $x = 8$, $z = 3$, being substituted in the equation,

$$6x + y + 5z = 68,$$

give, $\qquad 48 + y + 15 = 68,$

whence, $\qquad y = 5.$

Therefore, in order to form a pound of the fourth ingot, we must take 8 ounces of the first, 5 ounces of the second, and 3 of the third.

VERIFICATION.

If there be 7 ounces of silver in 16 ounces of the first ingot, in eight ounces of it there should be a number of ounces of silver expressed by

$$\frac{7 \times 8}{16}.$$

In like manner,

$$\frac{12 \times 5}{16}, \text{ and } \frac{4 \times 3}{16},$$

will express the quantity of silver contained in 5 ounces of the second ingot, and 3 ounces of the third.

Now, we have,

$$\frac{7 \times 8}{16} + \frac{12 \times 5}{16} + \frac{4 \times 3}{16} = \frac{128}{16} = 8;$$

therefore, a pound of the fourth ingot contains 8 ounces of silver, as required by the enunciation. The same conditions may be verified with respect to the copper and pewter.

3. *A's* age is double *B's*, and *B's* is triple of *C's*, and the sum of all their ages is 140: what is the age of each?

*Ans.* $A's = 84$; $B's = 42$; and $C's = 14.$

4. A person bought a chaise, horse, and harness, for £60; the horse came to twice the price of the harness, and the chaise to twice the cost of the horse and harness: what did he give for each?

*Ans.* $\begin{cases} £13 \;\; 6s. \;\; 8d. \;\; \text{for the horse.} \\ £6 \;\; 13s. \;\; 4d. \;\; \text{for the harness.} \\ £40 \;\;\;\;\;\;\;\;\;\;\; \text{for the chaise.} \end{cases}$

5. Divide the number 36 into three such parts that $\frac{1}{2}$ of the first, $\frac{1}{3}$ of the second, and $\frac{1}{4}$ of the third, may be all equal to each other.          *Ans.* 8, 12, and 16.

6. If *A* and *B* together can do a piece of work in 8 days, *A* and *C* together in 9 days, and *B* and *C* in ten days, how many days would it take each to perform the same work alone?          *Ans.* $A$, $14\frac{34}{49}$; $B$, $17\frac{23}{41}$; $C$, $23\frac{7}{31}$.

7. Three persons, *A*, *B*, and *C*, begin to play together, having among them all $600. At the end of the first game *A* has won one-half of *B's* money, which, added to his own, makes double the amount *B* had at first. In the second game, *A* loses and *B* wins just as much as *C* had at the beginning, when *A* leaves off with exactly what he had at first: how much had each at the beginning?

*Ans.* $A$, $300; $B$, $200; $C$ $100.

8. Three persons, *A*, *B*, and *C*, together possess $3640.

If $B$ gives $A$ $400 of his money, then $A$ will have $320 more than $B$; but if $B$ takes $140 of $C$'s money, then $B$ and $C$ will have equal sums : how much has each ?

Ans. $A$, $800 ; $B$, $1280 ; $C$, $1560.

9. Three persons have a bill to pay, which neither alone is able to discharge. $A$ says to $B$, "Give me the, 4th of your money, and then I can pay the bill." $B$ says to $C$, "Give me the 8th of yours, and I can pay it." But $C$ says to $A$, "You must give me the half of yours before I can pay it, as I have but $8 " : what was the amount of their bill, and how much money had $A$ and $B$?

Ans. $\begin{cases} \text{Amount of the bill, } \$13. \\ A \text{ had } \$10, \text{ and } B \ \$12. \end{cases}$

10. A person possessed a certain capital, which he placed out at a certain interest. Another person, who possessed 10000 dollars more than the first, and who put out his capital 1 per cent. more advantageously, had an annual income greater by 800 dollars. A third person, who possessed 15000 dollars more than the first, putting out his capital 2 per cent. more advantageously, had an annual income greater by 1500 dollars. Required, the capitals of the three persons, and the rates of interest.

Ans. $\begin{cases} \text{Sums at interest, } \$30000, \ \$40000, \ \$45000. \\ \text{Rates of interest, } \quad 4 \qquad 5 \qquad 6 \text{ pr. ct.} \end{cases}$

11. A widow receives an estate of $15000 from her deceased husband, with directions to divide it among two sons and three daughters, so that each son may receive twice as much as each daughter, and she herself to receive $1000 more than all the children together : what was her share, and what the share of each child?

Ans. $\begin{cases} \text{The widow's share, } \$8000 \\ \text{Each son's,} \qquad\qquad \$2000 \\ \text{Each daughter's,} \qquad \$1000 \end{cases}$

12. A certain sum of money is to be divided between three persons, *A*, *B*, and *C*. *A* is to receive $3000 less than half of it, *B* $1000 less than one-third part, and *C* to receive $800 more than the fourth part of the whole: what is the sum to be divided, and what does each receive?

$$Ans. \begin{cases} \text{Sum,} & \$38400. \\ A \text{ receives } \$16200. \\ B \quad `` \quad \$11800. \\ C \quad `` \quad \$10400. \end{cases}$$

13. A person has three horses, and a saddle which is worth $220. If the saddle be put on the back of the first horse, it will make his value equal to that of the second and third; if it be put on the back of the second, it will make his value double that of the first and third; if it be put on the back of the third, it will make his value triple that of the first and second: what is the value of each horse?

*Ans.* 1st, $20; 2d, $100; 3d, $140.

14. The crew of a ship consisted of her complement of sailors, and a number of soldiers. There were 22 sailors to every three guns, and 10 over; also, the whole number of hands was five times the number of soldiers and guns together. But after an engagement, in which the slain were one-fourth of the survivors, there wanted 5 men to make 13 men to every two guns: required, the number of guns, soldiers and sailors.

*Ans.* 90 guns, 55 soldiers, and 670 sailors.

15. Three persons have $96, which they wish to divide equally between them. In order to do this, *A*, who has the most, gives to *B* and *C* as much as they have already; then *B* divides with *A* and *C* in the same manner, that is, by giving to each as much as he had after *A* had divided with them. *C* then makes a division with *A* and *B*, when it is

found that they all have equal sums: how much had each
at first? *Ans.* 1st, $52; 2d, $28; 3d, $16.

16. Divide the number $a$ into three such parts, that the
first shall be to the second as $m$ to $n$, and the second to the
third as $p$ to $q$.

$$x = \frac{amp}{mp+np+nq}, \quad y = \frac{anp}{mp+np+nq}, \quad z = \frac{anq}{mp+np+nq}.$$

17. Three masons, $A$, $B$, and $C$, are to build a wall. $A$
and $B$ together can do it in 12 days; $B$ and $C$ in 20 days;
and $A$ and $C$ in 15 days: in what time can each do it alone,
and in what time can they all do it if they work together?
*Ans.* $A$, in 20 days; $B$, in 30; and $C$, in 60; all, in 10.

# CHAPTER VI.

### FORMATION OF POWERS.

**118.** A POWER of a quantity is the product obtained by taking that quantity any number of times as a factor.

If the quantity be taken once as a factor, we have the first power; if taken twice, we have the second power; if three times, the third power; if $n$ times, the $n^{th}$ power, $n$ being any whole number whatever.

A power is indicated by means of the exponential sign thus,

$$a = a^1 \quad \text{denotes first power of } a.*$$
$$a \times a = a^2 \quad \text{``} \quad \text{square, or 2d power of } a.$$
$$a \times a \times a = a^3 \quad \text{``} \quad \text{cube, or third power of } a.$$
$$a \times a \times a \times a = a^4 \quad \text{``} \quad \text{fourth power of } a.$$
$$a \times a \times a \times a \times a = a^5 \quad \text{``} \quad \text{fifth power of } a.$$
$$a \times a \times a \times a .... = a^m \quad \text{``} \quad m^{th} \text{ power of } a.$$

In every power there are three things to be considered:

1st. The quantity which enters as a factor, and which is called the first power.

2d. The small figure which is placed at the right, and a little above the letter, is called the *exponent* of the

* Since $a^0 = 1$ (Art. 49), $a^0 \times a = 1 \times a = a^1$; so that the two factors of $a^1$, are 1 and $a$.

---

118. What is a power of a quantity? What is the power when the quantity is taken once as a factor? When taken twice? Three times? $n$ times? How is a power indicated? In every power, how many things are considered? Name them.

power, and shows how many times 'the letter enters as a factor.

3d. The power itself, which is the final product, or result of the multiplications.

## POWERS OF MONOMIALS.

**119.** Let it be required to raise the monomial $2a^3b^2$ to the fourth power. We have,

$$(2a^3b^2)^4 = 2a^3b^2 \times 2a^3b^2 \times 2a^3b^2 \times 2a^3b^2,$$

which merely expresses that the fourth power is equal to the product which arises from taking the quantity four times as a factor. By the rules for multiplication, this product is

$$(2a^3b^2)^4 = 2^4 a^{3+3+3+3} b^{2+2+2+2} = 2^4 a^{12} b^8;$$

from which we see,

1st. That the coefficient 2 must be raised to the 4th power; and,

2d. That the exponent of each letter must be multiplied by 4, the exponent of the power.

As the same reasoning applies to every example, we have, for the raising of monomials to any power, the following

### RULE.

I. *Raise the coefficient to the required power :*

II. *Multiply the exponent of each letter by the exponent of the power.*

### EXAMPLES.

1. What is the square of $3a^2y^3$?  *Ans.* $9a^4y^6$.

---

119. What is the rule for raising a monomial to any power? When the monomial is positive, what will be the sign of its powers? When negative, what powers will be plus? what minus?

2. What is the cube of $6a^5y^2x$?     $Ans.$ $216a^{15}y^6x^3$.

3. What is the fourth power of $2a^3y^3b^\circ$?     $16a^{12}y^{12}b^{20}$.

4. What is the square of $a^2b^5y^3$?     $Ans.$ $a^4b^{10}y^6$.

5. What is the seventh power of $a^2bcd^3$?
$Ans.$ $a^{14}b^7c^7d^{21}$.

6. What is the sixth power of $a^2b^3c^2d$?
$Ans.$ $a^{12}b^{18}c^{12}d^6$.

7. What is the square and cube of $-2a^2b^2$?

| Square. | Cube. |
|---|---|
| $-2a^2b^2$ | $-2a^2b^2$ |
| $-2a^2b^2$ | $-2a^2b^2$ |
| $+4a^4b^4.$ | $+4a^4b^4$ |
|  | $-2a^2b^2$ |
|  | $-8a^6b^6.$ |

By observing the way in which the powers are formed, we may conclude,

1st. *When the monomial is positive, all the powers will be positive.*

2d. *When the monomial is negative, all even powers will be positive, and all odd will be negative.*

8. What is the square of $-2a^4b^5$?     $Ans.$ $4a^8b^{10}$.

9. What is the cube of $-5a^nb^2$?     $Ans.$ $-125a^{3n}b^6$.

10. What is the eighth power of $-a^3xy^2$?
$Ans.$ $+a^{24}x^8y^{16}$.

11. What is the seventh power of $-a^mb^nc$?
$Ans.$ $-a^{7m}b^{7n}c^7$.

12. What is the sixth power of $2ab^6y^5$?
$Ans.$ $64a^6b^{36}y^{30}$.

13. What is the ninth power of $- a^n bc^2$ ?

$$Ans. \quad - a^{9n}b^9c^{18}.$$

14. What is the sixth power of $- 3ab^2d$ ?

$$Ans. \quad 729a^6b^{12}d^6.$$

15. What is the square of $- 10a^m b^n c^3$ ?

$$Ans. \quad 100a^{2m}b^{2n}c^6.$$

16. What is the cube of $- 9a^m b^n d^3 f^2$ ?

$$Ans. \quad - 729a^{3m}b^{3n}d^9f^6.$$

17. What is the fourth power of $- 4a^5b^3c^4d^5$ ?

$$Ans. \quad 256a^{20}b^{12}c^{16}d^{20}.$$

18. What is the cube of $- 4a^{2m}b^{2n}c^3d$ ?

$$Ans. \quad - 64a^{6m}b^{6n}c^9d^3.$$

19. What is the fifth power of $2a^3b^2xy$ ?

$$Ans. \quad 32a^{15}b^{10}x^5y^5.$$

20. What is the square of $20x^n y^m c^5$ ? $Ans. \quad 400x^{2n}y^{2m}c^{10}.$

21. What is the fourth power of $3a^n b^{2n} c^3$ ?

$$Ans. \quad 81a^{4n}b^{8n}c^{12}.$$

22. What is the fifth power of $- c^n d^{3m} x^2 y^2$ ?

$$Ans. \quad - c^{5n}d^{15m}x^{10}y^{10}.$$

23. What is the sixth power of $- a^n b^{2n} c^m$ ?

$$Ans. \quad a^{6n}b^{12n}c^{6m}.$$

24. What is the fourth power of $- 2a^2c^2d^3$.

$$Ans. \quad 16a^8c^8d^{12}.$$

## POWERS OF FRACTIONS.

**120.** From the definition of a power, and the rule for the multiplication of fractions, the cube of the fraction $\frac{a}{b}$, is written,

$$\left(\frac{a}{b}\right)^3 = \frac{a}{b} \times \frac{a}{b} \times \frac{a}{b} = \frac{a^3}{b^3};$$

120. What is the rule for raising a fraction to any power?

and since any fraction raised to any power, may be written under the same form, we find any power of a fraction by the following

RULE.

*Raise the numerator to the required power for a new numerator, and the denominator to the required power for a new denominator.*

The rule for signs is the same as in the last article.

EXAMPLES

Find the powers of the following fractions:

1. $\left(\dfrac{a-c}{b+c}\right)^2$.     *Ans.* $\dfrac{a^2-2ac+c^2}{b^2+2bc+c^2}$.

2. $\left(\dfrac{xy}{3bc}\right)^3$.     *Ans.* $\dfrac{x^3y^3}{27b^3c^3}$.

3. $\left(\dfrac{-x^2y}{2ab}\right)^4$.     *Ans.* $\dfrac{x^8y^4}{16a^4b^4}$.

4. $\left(\dfrac{2ax^2y}{3bc^2}\right)^2$.     *Ans.* $\dfrac{4a^2x^4y^2}{9b^2c^4}$.

5. $\left(-\dfrac{dx}{3y^2}\right)^3$.     *Ans.* $-\dfrac{d^3x^3}{27y^6}$.

6. $\left(\dfrac{axy^3}{2bz^2}\right)^3$.     *Ans.* $\dfrac{a^3x^3y^9}{8b^3z^6}$.

7. $\left(-\dfrac{3ay^4}{2b^2x}\right)^4$.     *Ans.* $\dfrac{81a^4y^{16}}{16b^8x^4}$.

8. Fourth power of $\dfrac{ab^2c}{2x^2y^2}$.     *Ans.* $\dfrac{a^4b^8c^4}{16x^8y^8}$.

9. Cube of $\dfrac{x-y}{x+y}$.     *Ans.* $\dfrac{x^3-3x^2y+3xy^2-y^3}{x^3+3x^2y+3xy^2+y^3}$.

10. Fourth power of $\dfrac{2a^m x^n}{4a^p y^q}$.     *Ans.* $\dfrac{a^{4m} x^{4n}}{16a^{4p} y^{4q}}$.

11. Fifth power of $-\dfrac{9bc^n x^m}{18y^p z^q}$.     *Ans.* $-\dfrac{b^5 c^{5n} x^{5m}}{32y^{5p} z^{5q}}$.

## POWERS OF BINOMIALS.

**121.** A Binomial, like a monomial, may be raised to any power by the process of continued multiplication.

1. Find the fifth power of the binomial $a + b$.

$a + b$ . . . . . . . . . . . . 1st power.
$a + b$
————
$a^2 + ab$
$\phantom{a^2} + ab + b^2$
————
$a^2 + 2ab + b^2$ . . . . . . . . 2d power.
$a + b$
————
$a^3 + 2a^2 b + \phantom{2}ab^2$
$\phantom{a^3} + \phantom{2}a^2 b + 2ab^2 + b^3$
————
$a^3 + 3a^2 b + 3ab^2 + b^3$ . . . . 3d power.
$a + b$
————
$a^4 + 3a^3 b + 3a^2 b^2 + \phantom{3}ab^3$
$\phantom{a^4} + \phantom{3}a^3 b + 3a^2 b^2 + 3ab^3 + b^4$
————
$a^4 + 4a^3 b + 6a^2 b^2 + 4ab^3 + b^4$    4th power.
$a + b$
————
$a^5 + 4a^4 b + 6a^3 b^2 + 4a^2 b^3 + \phantom{4}ab^4$
$\phantom{a^5} + \phantom{4}a^4 b + 4a^3 b^2 + 6a^2 b^3 + 4ab^4 + b^5$
————
$a^5 + 5a^4 b + 10a^3 b^2 + 10a^2 b^3 + 5ab^4 + b^5$    *Ans.*

---

121. How may a binomial be raised to any power?

122. How does the number of multiplications compare with the exponent of the power? If the exponent is 4, what is the number of multiplications? How many when it is $m$? How many things are considered in the raising of powers? Name them.

Note.—**122.** It will be observed that the number of multiplications is always 1 less than the units in the exponent of the power. Thus, if the exponent is 1, no multiplication is necessary. If it is 2, we multiply once; if it is 3, twice; if 4, three times, &c. The powers of polynomials may be expressed by means of an exponent. Thus, to express that $a + b$ is to be raised to the 5th power, we write

$$(a + b)^5;$$

if to the $m$th power, we write

$$(a + b)^m.$$

2. Find the 5th power of the binomial $a - b$.

$$
\begin{array}{l}
a - b \quad \ldots \ldots \ldots \ldots \text{ 1st power.} \\
a - b \\
\hline
a^2 - ab \\
\quad - ab + b^2 \\
\hline
a^2 - 2ab + b^2 \quad \ldots \ldots \ldots \text{ 2d power.} \\
a - b \\
\hline
a^3 - 2a^2b + ab^2 \\
\quad - a^2b + 2ab^2 - b^3 \\
\hline
a^3 - 3a^2b + 3ab^2 - b^3 \quad \ldots \ldots \text{ 3d power.} \\
a - b \\
\hline
a^4 - 3a^3b + 3a^2b^2 - ab^3 \\
\quad - a^3b + 3a^2b^2 - 3ab^3 + b^4 \\
\hline
a^4 - 4a^3b + 6a^2b^2 - 4ab^3 + b^4 \quad . \text{ 4th power.} \\
a - b \\
\hline
a^5 - 4a^4b + 6a^3b^2 - 4a^2b^3 + ab^4 \\
\quad - a^4b + 4a^3b^2 - 6a^2b^3 + 4ab^4 - b^5 \\
\hline
a^5 - 5a^4b + 10a^3b^2 - 10a^2b^3 + 5ab^4 - b^5 \quad \textit{Ans.}
\end{array}
$$

In the same way the higher powers may be obtained. By examining the powers of these binomials, it is plain that four things must be considered:

1st. The number of terms of the power.
2d. The signs of the terms.
3d. The exponents of the letters.
4th. The coefficients of the terms.

Let us see according to what laws these are formed.

## Of the Terms.

**123.** By examining the several multiplications, we shall observe that the first power of a binomial contains two terms; the second power, three terms; the third power, four terms; the fourth power, five; the fifth power, six, &c.; and hence we may conclude:

*That the number of terms in any power of a binomial, is greater by one than the exponent of the power.*

## Of the Signs of the Terms.

**124.** It is evident that when both terms of the given binomial are plus, *all the terms of the power will be plus.*

If the second term of the binomial is negative, then *all the odd terms, counted from the left, will be positive, and all the even terms negative.*

---

123. How many terms does the first power of a binomial contain? The second? The third? The *n*th power?
· 124. If both terms of a binomial are positive, what will be the signs of the terms of the power? If the second term is negative, how are the signs of the terms?

## *Of the Exponents.*

**125.** The letter which occupies the first place in a binomial, is called the *leading letter.* Thus, $a$ is the leading letter in the binomials $a + b$, and $a - b$.

1st. It is evident that the exponent of the leading letter in the first term, will be the same as the exponent of the power; and that this exponent will diminish by one in each term to the right, until we reach the last term, when it will be 0 (Art. 49).

2d. The exponent of the second letter is 0 in the first term, and increases by one in each term to the right, to the last term, when the exponent is the same as that of the given power.

3d. The sum of the exponents of the two letters, in any term, is equal to the exponent of the given power. This last remark will enable us to verify any result obtained by means of the binomial formula.

Let us now apply these principles in the two following examples, in which the coefficients are omitted :

$$(a + b)^6 \ldots a^6 + a^5b + a^4b^2 + a^3b^3 + a^2b^4 + ab^5 + b^6,$$
$$(a - b)^6 \ldots a^6 - a^5b + a^4b^2 - a^3b^3 + a^2b^4 - ab^5 + b^6.$$

As the pupil should be practised in writing the terms with their proper signs, without the coefficients, we will add a few more examples.

---

125. Which is the leading letter of a binomial ? What is the exponent of this letter in the first term ? How does it change in the terms towards the right ? What is the exponent of the second letter in the second term ? How does it change in the terms towards the right ? What is it in the last term ? What is the sum of the exponents in any term equal to ?

1. $(a + b)^3$ . . $a^3 + a^2b + ab^2 .+ b^3$.
2. $(a - b)^4$ . . $a^4 - a^3b + a^2b^2 - ab^3 + b^4$.
3. $(a + b)^5$ . . $a^5 + a^4b + a^3b^2 + a^2b^3 + ab^4 + b^5$.
4. $(a - b)^7$ . . $a^7 - a^6b + a^5b^2 - a^4b^3 + a^3b^4 - a^2b^5 + ab^6 - b^7$.

## Of the Coefficients.

**126.** The coefficient of the first term is 1. The coeffi cient of the second term is the same as the exponent of the given power. The coefficient of the third term is found by multiplying the coefficient of the second term by the expo nent of the leading letter in that term, and dividing the product by 2. And finally:

*If the coefficient of any term be multiplied by the expo nent of the leading letter in that term, and the product divided by the number which marks the place of the term from the left, the quotient will be the coefficient of the next term.*

Thus, to find the coefficients in the example,

$(a - b)^7$ . . . $a^7 - a^6b + a^5b^2 - a^4b^3 + a^3b^4 - a^2b^5 + ab^6 - b^7$,

we first place the exponent 7 as a coefficient of the second term. Then, to find the coefficient of the third term, we multiply 7 by 6, the exponent of $a$, and divide by 2. The quotient, 21, is the coefficient of the third term. To find the coefficient of the fourth, we multiply 21 by 5, and divide the product by 3; this gives 35. To find the coefficient of the fifth term, we multiply 35 by 4, and divide the product by 4; this gives 35. The coefficient of the sixth term, found

126. What is the coefficient of the first term? What is the coefficient of the second term? How do you find the coefficient of the third term? How do you find the coefficient of any term? What are the coefficients of the first and last terms? How are the coefficients of the exponents of any two terms equally distant from the two extremes?

8

in the same way, is 21; that of the seventh, 7; and that of the eighth, 1. Collecting these coefficients,

$$(a - b)^7 =$$
$$a^7 - 7a^6b + 21a^5b^2 - 35a^4b^3 + 35a^3b^4 - 21a^2b^5 + 7ab^6 - b^7.$$

NOTE.—We see, in examining this last result, that the *coefficients of the extreme terms are each* 1, *and that the coefficients of terms equally distant from the extreme terms are equal.* It will, therefore, be sufficient to find the coefficients of the first half of the terms, and from these the others may be immediately written.

### EXAMPLES

1. Find the fourth power of $a + b$.
   Ans. $a^4 + 4a^3b + 6a^2b^2 + 4ab^3 + b^4$.

2. Find the fourth power of $a - b$.
   Ans. $a^4 - 4a^3b + 6a^2b^2 - 4ab^3 + b^4$.

3. Find the fifth power of $a + b$.
   Ans. $a^5 + 5a^4b + 10a^3b^2 + 10a^2b^3 + 5ab^4 + b^5$.

4. Find the fifth power of $a - b$.
   Ans. $a^5 - 5a^4b + 10a^3b^2 - 10a^2b^3 + 5ab^4 - b^5$.

5. Find the sixth power of $a + b$.
   $a^6 + 6a^5b + 15a^4b^2 + 20a^3b^3 + 15a^2b^4 + 6ab^5 + b^6$.

6. Find the sixth power of $a - b$.
   $a^6 - 6a^5b + 15a^4b^2 - 20a^3b^3 + 15a^2b^4 - 6ab^5 + b^6$.

**127.** When the terms of the binomial have coefficients, we may still write out any power of it by means of the Binomial Formula.

7. Let it be required to find the cube of $2c + 3d$.

$$(a + b)^3 = a^3 + 3a^2b + 3ab^2 + b^3.$$

Here, $2c$ takes the place of $a$ in the formula, and $3d$ the place of $b$. Hence, we have,

$$(2c+3d)^3 = (2c)^3+3.(2c)^2.3d+3(2c)(3d)^2+(3d)^3 \quad . \quad (1.)$$

and by performing the indicated operations, we have,

$$(2c + 3d)^3 = 8c^3 + 36c^2d + 54cd^2 + 27d^3.$$

If we examine the second member of Equation ( 1 ), we see that each term is made up of three factors: 1st, the numerical factor; 2d, some power of $2c$; and 3d, some power of $3d$. The powers of $2c$ are arranged in descending order towards the right, the last term involving the 0 power of $2c$ or 1; the powers of $3d$ are arranged in ascending order from the first term, where the 0 power enters, to the last term.

The operation of raising a binomial involving coefficients, is most readily effected by writing the three factors of each term in a vertical column, and then performing the multiplications as indicated below.

Find, by this method, the cube of $2c + 3d$.

OPERATION.

| 1 | + 3 | + 3 | + 1 | Coefficients. |
|---|---|---|---|---|
| $8c^3 +$ | $4c^2$ | $+ 2c$ | $+ 1$ | Powers of $2c$ |
| 1 | $+ 3d$ | $+ 9d^2$ | $+ 27d^3$ | Powers of $3d$ |

$$(2c + d)^3 = 8c^3 + 36c^2d + 54cd^2 + 27d^3$$

The preceding operation hardly requires explanation. In the first line, write the numerical coefficients corresponding to the particular power; in the second line, write the descending powers of the leading term to the 0 power; in the third line, write the ascending powers of the following term from the 0 power upwards. It will be easiest to commence

the second line on the right hand. The multiplication should
be performed from above, downwards.

8. Find the 4th power of $3a^2c - 2bd$.

$$(a + b)^4 = a^4 + 4a^3b + 6a^2b^2 + 4ab^3 + b^4.$$

| 1 | $+$ 4 | $+$ 6 | $+$ 4 | $+$ 1 |
|---|---|---|---|---|
| $81a^8c^4 +$ | $27a^6c^3 +$ | $9a^4c^2 +$ | $3a^2c +$ | $1$ |
| $1 -$ | $2bd +$ | $4b^2d^2 -$ | $8b^3d^3$ | $+ 16b^4d$ . |

$$81a^8c^4 - 216a^6c^2bd + 216a^4c^2b^2d^2 - 96a^2cb^3d^3 + 16b^4d^4. \ast$$

9. What is the cube of $3x - 6y$ ?
$\qquad$ *Ans.* $27x^3 - 162x^2y + 324xy^2 - 216y^3.$

10. What is the fourth power of $a - 3b$?
$\qquad$ *Ans.* $a^4 - 12a^3b + 54a^2b^2 - 108ab^3 + 81b^4.$

11. What is the fifth power of $c - 2d$?
$\qquad$ *Ans.* $c^5 - 10c^4d + 40c^3d^2 - 80c^2d^3 + 80cd^4 - 32d^5.$

12. What is the cube of $5a - 3d$?
$\qquad$ *Ans.* $125a^3 - 225a^2d + 135ad^2 - 27d^3.$

* This ingenious method of writing the development of a binomial is due to
Professor WILLIAM G. PECK, of Columbia College.

## CHAPTER VII.

SQUARE ROOT. RADICALS OF THE SECOND
DEGREE.

**128.** The Square Root of a number is one of its two
equal factors. Thus, $6 \times 6 = 36$; therefore, 6 is the square
root of 36.

The symbol for the square root, is $\sqrt{\phantom{x}}$ , or the fractional
exponent $\frac{1}{2}$; thus,

$$\sqrt{a}, \quad \text{or} \quad a^{\frac{1}{2}},$$

indicates the square root of $a$, or that one of the two equal
factors of $a$ is to be found. The operation of finding such
factor is called, *Extracting the Square Root.*

**129.** Any number which can be resolved into *two equal
integral factors*, is called a *perfect square.*

The following Table, verified by actual multiplication, indicates all the perfect squares between 1 and 100.

TABLE.

| 1, | 4, | 9, | 16, | 25, | 36, | 49, | 64, | 81, | 100, | squares. |
|----|----|----|-----|-----|-----|-----|-----|-----|------|----------|
| 1, | 2, | 3, | 4, | 5, | 6, | 7, | 8, | 9, | 10, | roots. |

128. What is the square root of a number? Wha is the operation of
finding the equal factor called?
129. What is a perfect square? How many perfect squares are there
between 1 and 100, including both numbers? What are they?

We may employ this table for finding the square root of any perfect square between 1 and 100.

*Look for the number in the first line; if it is found there, its square root will be found immediately under it.*

If the given number is less than 100, and not a perfect square, *it will fall between two numbers of the upper line, and its square root will be found between the two numbers directly below; the lesser of the two will be the entire part of the root, and will be the true root to within less than 1.*

Thus, if the given number is 55, it is found between the perfect squares 49 and 64, and its root is 7 and a decimal fraction.

NOTE.—There are ten perfect squares between 1 and 100, if we include both numbers; and eight, if we exclude both.

If a number is greater than 100, its square root will be greater than 10, that is, it will contain *tens* and *units*. Let $N$ denote such a number, $x$ the tens of its square root, and $y$ the units; then will,

$$N = (x + y)^2 = x^2 + 2xy + y^2 = x^2 + (2x + y)y.$$

That is, the number is equal to the *square of the tens* in its roots, plus *twice the product of the tens by the units*, plus *the square of the units*.

EXAMPLE.

1. Extract the square root of 6084.

Since this number is composed of more than two places of figures, its root will contain more than one.   But since it is less than 10000, which    $60\ 84$
is the square of 100, the root will contain but two figures; that is, units and tens.

Now, the square of the tens must be found in the two left-hand figures, which we will separate from the other two by putting a point over the place of units, and a second over the place of hundreds. These parts, of two figures each, are called *periods*. The part 60 is comprised between the two squares 49 and 64, of which the roots are 7 and 8; hence, 7 *expresses the number of tens sought;* and the required root is composed of 7 tens and a certain number of units.

The figure 7 being found, we write it on the right of the given number, from which we separate it by a vertical line : then we subtract its square, 49, from 60, which leaves a remainder of 11, to which we bring down the two

$$
\begin{array}{r|l}
\dot{6}0\ \dot{8}4 & 7\dot{8} \\
49 & \\
\hline
7 \times 2 = 14\ 8 \,|\, 118\ 4 \\
118\ 4 \\
\hline
0
\end{array}
$$

next figures, 84. The result of this operation, 1184, contains *twice the product of the tens by the units, plus the square of the units.*

But since tens multiplied by units cannot give a product of a less unit than tens, it follows that the last figure, 4, can form no part of the double product of the tens by the units; this double product is therefore found in the part 118, which we separate from the units' place, 4.

Now if we double the tens, which gives 14, and then divide 118 by 14, the quotient 8 *will express the units,* or a number greater than the units. This quotient can never be too small, since the part 118 will be at least equal to twice the product of the tens by the units; but it may be too large, for the 118, besides the double product of the tens by the units, may likewise contain tens arising from the square of the units. To ascertain if the quotient 8 expresses the right number of units, we write the 8 on the right of the 14, which gives 148, and then we multiply 148 by 8. This multiplication being effected, gives for a product, 1184, a number equal to the result of the first operation. Having subtracted the product, we find the remainder equal to 0; hence, 78 is the root required. In this operation, we form, 1st, the square of the tens; 2nd, the double product ot the tens by the units; and 3d, the square of the units.

Indeed, in the operations, we have merely subtracted from the given number 6084 : 1st, the square of 7 tens, or of 70; 2d, twice the product of 70 by 8; and, 3d, the square of 8; that is, the three parts which enter into the composition of the square, 70 + 8, or 78; and since the result of the subtraction is 0, it follows that 78 is the square root of 6084.

**130.** The operations in the last example have been performed on but two periods, but it is plain that the same methods of reasoning are equally applicable to larger numbers, for by changing the order of the units, we do not change the relation in which they stand to each other.

Thus, in the number 60 84 95, the two periods 60 84, have the same relation to each other as in the number 60 84; and hence the methods used in the last example are equally applicable to larger numbers.

**131.** Hence, for the extraction of the square root of numbers, we have the following

### RULE.

I. *Point off the given number into periods of two figures each, beginning at the right hand:* .

II. *Note the greatest perfect square in the first period on the left, and place its root on the right, after the manner of*

---

131. Give the rule for the extraction of the square root of numbers? What is the first step? What the second? What the third? What the fourth? What the fifth?

*a quotient in division ; then subtract the square of this root from the first period, and bring down the second period for a remainder :*

III. *Double the root already found, and place the result on the left for a divisor. Seek how many times the divisor is contained in the remainder, exclusive of the right-hand figure, and place the figure in the root and also at the right of the divisor :*

IV. *Multiply the divisor, thus augmented, by the last figure of the root, and subtract the product from the remainder, and bring down the next period for a new remainder. But if any of the products should be greater than the remainder, diminish the last figure of the root by one :*

V. *Double the whole root already found, for a new divisor, and continue the operation as before, until all the periods are brought down.*

**132.** NOTE.—1. If, after all the periods are brought down, there is no remainder, the given number is a perfect square.

2. The number of places of figures in the root will always be equal to the number of periods into which the given number is divided.

3. If the given number has not an exact root, there will be a remainder after all the periods are brought down, in which case ciphers may be annexed, forming new periods, for each of which there will be one decimal place in the root.

---

132. What takes place when the given number is a perfect square? How many places of figures will there be in the root? If the given number is not a perfect square, what may be done after all the periods are brought down?

1. What is the square root of 36729?

```
                          3̇ 67 2̇9 | 191.64+
                          1        |
                        ─────────────────────
                     2 9 | 267
                         | 261
                         |
                    38 1 | 629
                         | 381
                         |
                   382 6 | 24800
                         | 22956
                         |
                  3832 4 | 184400
                         | 153296
                         ─────────────────
                           31104  Rem.
```

In this example there are two periods of decimals, and, hence, two places of decimals in the root.

2. To find the square root of 7225.          *Ans.* 85.
3. To find the square root of 17689.          *Ans.* 133.
4. To find the square root of 994009.          *Ans.* 997.
5. To find the square root of 85673536.          *Ans.* 9256.
6. To find the square root of 67798756.          *Ans.* 8234.
7. To find the square root of 978121.          *Ans.* 989.
8. To find the square root of 956484.          *Ans.* 978.
9. What is the square root of 36372961? ·    *Ans.* 6031.
10. What is the square root of 22071204?    *Ans.* 4698.
11. What is the square root of 106929?          *Ans.* 327.
12. What of 12088868379025?          *Ans.* 3476905.
13. What of 2268741?          *Ans.* 1506.23 +.
14. What of 7596796?          *Ans.* 2756.22 +.
15. What is the square root of 96?    *Ans.* 9.79795 +.
16. What is the square root of 153?    *Ans.* 12.36931 +.
17. What is the square root of 101?    *Ans.* 10.04987 +.

18. What of 285970396044 ?    *Ans.* 534762.

19. What of 41605800625 ?    *Ans.* 203975.

20. What of 48303584206084 ?    *Ans.* 6950078.

EXTRACTION OF THE SQUARE ROOT OF FRACTIONS.

**133.** Since the square or second power of a fraction is obtained by squaring the numerator and denominator separately, it follows that

*The square root of a fraction will be equal to the square root of the numerator divided by the square root of the denominator.*

For example, the square root of $\frac{a^2}{b^2}$ is equal to $\frac{a}{b}$: for,

$$\frac{a}{b} \times \frac{a}{b} = \frac{a^2}{b^2}.$$

1. What is the square root of $\frac{1}{4}$?    *Ans.* $\frac{1}{2}$.

2. What is the square root of $\frac{9}{16}$?    *Ans.* $\frac{3}{4}$.

3. What is the square root of $\frac{64}{81}$?    *Ans.* $\frac{8}{9}$.

4. What is the square root of $\frac{256}{361}$?    *Ans.* $\frac{16}{19}$.

5. What is the square root of $\frac{16}{64}$?    *Ans.* $\frac{1}{2}$.

6. What is the square root of $\frac{4096}{61009}$?    *Ans.* $\frac{64}{247}$.

7. What is the square root of $\frac{582169}{956484}$?    *Ans.* $\frac{763}{978}$.

---

133. To what is the square root of a fraction equal?

**134.** If the numerator and denominator are not perfect squares, the root of the fraction cannot be exactly found. We can, however, easily find the approximate root.

### RULE.

*Multiply both terms of the fraction by the denominator: Then extract the square root of the numerator, and divide this root by the root of the denominator; the quotient will be the approximate root.*

1. Find the square root of $\frac{3}{5}$.

Multiplying the numerator and denominator by 5

$$\sqrt{\frac{3}{5}} = \sqrt{\frac{15}{25}} = \frac{\sqrt{15}}{5} = (3.8729 \,+) \div 5 ;$$

hence,   $(3.8729 \,+) \div 5 = .7745 \,+ = Ans.$

2. What is the square root of $\frac{7}{4}$ ?        *Ans.* 1.32287 +.

3. What is the square root of $\frac{14}{9}$ ?        *Ans.* 1.24721 +.

4. What is the square root of $11\frac{11}{16}$ ?        *Ans.* 3.41869 +.

5. What is the square root of $7\frac{13}{36}$ ?        *Ans.* 2.71313 +.

6. What is the square root of $8\frac{15}{49}$ ?        *Ans.* 2.88203 +.

7. What is the square root of $\frac{5}{12}$ ?        *Ans.* 0.64549 +.

8. What is the square root of $10\frac{3}{10}$ ?        *Ans.* 3.20936 +.

---

134. What is the rule when the numerator and denominator are not perfect squares?

**135.** Finally, instead of the last method, we may, if we please,

*Change the common fraction into a decimal, and continue the division until the number of decimal places is double the number of places required in the root. Then extract the root of the decimal by the last rule.*

EXAMPLES.

1. Extract the square of $\frac{11}{14}$ to within .001. This number, reduced to decimals, is 0.785714 to within 0.000001 ; but the root of 0.785714 to the nearest unit, is .886; hence, 0.886 is the root of $\frac{11}{14}$ to within .001.

2. Find the $\sqrt{2\frac{13}{15}}$ to within 0.0001. *Ans.* 1.6931 +.

3. What is the square root of $\frac{1}{17}$ ? *Ans.* 0.24253 +.

4. What is the square root of $\frac{7}{8}$ ? *Ans.* 0.93541 +.

5. What is the square root of $\frac{5}{3}$ ? *Ans.* 1.29099 +.

EXTRACTION OF THE SQUARE ROOT OF MONOMIALS.

**136.** In order to discover the process for extracting the square root of a monomial, we must see how its square is formed.

By the rule for the multiplication of monomials (Art. 42), we have,

$$(5a^2b^3c)^2 = 5a^2b^3c \times 5a^2b^3c = 25a^4b^6c^2;$$

135. What is a second method of finding the approximate root?
136. Give the rule for extracting the square root of monomials?

that is, in order to square a monomial, it is necessary to *square its coefficient and double the exponent of each of the letters.* Hence, to find the square root of a monomial, we have the following

### RULE.

I. *Extract the square root of the coefficient for a new coefficient:*

II. *Divide the exponent of each letter by* 2, *and then annex all the letters with their new exponents.*

Since like signs in two factors give a plus sign in the product, the square of $-a$, as well as that of $+a$, will be $+a^2$; hence, the square root of $a^2$ is either $+a$, or $-a$. Also, the square root of $25a^2b^4$, is either $+5ab^2$, or $-5ab^2$. Whence we conclude, that if a monomial is positive, its square root may be affected either with the sign $+$ or $-$; thus, $\sqrt{9a^4} = \pm 3a^2$; for, $+3a^2$ or $-3a^2$, squared, gives $+9a^4$. The double sign $\pm$, with which the root is affected, is read *plus and minus.*

### EXAMPLES.

1. What is the square root of $64a^6b^4$?

$$\sqrt{64a^6b^4} = +8a^3b^2; \text{ for } +8a^3b^2 \times +8a^3b^2 = +64a^6b^4$$

and, $\sqrt{64a^6b^4} = -8a^3b^2$; for $-8a^3b^2 \times -8a^3b^2 = +64a^6b^4$

Hence, $\qquad \sqrt{64a^6b^4} = \pm 8a^3b^4.$

2. Find the square root of $625a^2b^8c^6$. $\qquad \pm 25ab^4c^3.$

3. Find the square root of $576a^4b^6c^8$. $\qquad \pm 24a^2b^3c^4.$

4. Find the square root of $196x^6y^2z^4$. $\qquad \pm 14x^3yz^2.$

5. Find the square root of $441a^8b^6c^{10}d^{16}$. $\quad \pm 21a^4b^3c^5d^8.$

6. Find the square root of $784a^{12}b^{14}c^{16}d^2$. $\quad \pm 28a^6b^7c^8d.$

7. Find the square root of $81a^8b^4c^6$. $\qquad \pm 9a^4b^2c^3.$

Notes.—**137.** 1. From the preceding rule it follows, that when a monomial is a perfect square, *its numerical coefficient is a perfect square, and all its exponents even numbers.* Thus, $25a^4b^2$ is a perfect square.

2. If the proposed monomial were *negative*, it would be impossible to extract its square root, since it has just been shown (Art. 136) that the square of every quantity, whether positive or negative, is essentially positive. Therefore,

$$\sqrt{-9}, \quad \sqrt{-4a^2}, \quad \sqrt{-8a^2b},$$

are algebraic symbols which indicate operations that cannot be performed. They are called *imaginary quantities*, or rather, *imaginary expressions*, and are frequently met with in the resolution of equations of the second degree.

### IMPERFECT SQUARES.

**138.** When the coefficient is not a *perfect square*, or when the exponent of any letter is *uneven*, the monomial is an *imperfect square :* thus, $98ab^4$ is an *imperfect square.* Its root is then indicated by means of the radical sign; thus,

$$\sqrt{98ab^4}.$$

Such quantities are called, *radical quantities*, or *radicals of the second degree :* hence,

A radical quantity, is the indicated root of an imperfect power.

---

137. When is a monomial a perfect square? What monomials are these whose square roots cannot be extracted? What are such expressions called?

138. When is a monomial an imperfect square? What are such quantities called? What is a radical quantity?

## TRANSFORMATION OF RADICALS.

**139.** Let $a$ and $b$ denote any two numbers, and $p$ the product of their square roots: then,

$$\sqrt{a} \times \sqrt{b} = p \quad \ldots \ldots \quad (1.)$$

Squaring both members, we have,

$$a \times b = p^2 \quad \ldots \ldots \quad (2.)$$

Then, extracting the square root of both members of (2),

$$\sqrt{ab} = p \quad \ldots \ldots \quad (3.)$$

And since the second members are the same in Equations (1) and (3), the first members are equal: that is,

*The square root of the product of two quantities is equal to the product of their square roots.*

**140.** Let $a$ and $b$ denote any two numbers, and $q$ the quotient of their square roots; then,

$$\frac{\sqrt{a}}{\sqrt{b}} = q \quad \ldots \ldots \quad (1.)$$

Squaring both members, we have,

$$\frac{a}{b} = q^2 \quad \ldots \ldots \quad (2.)$$

then extracting the square root of both members of (2),

$$\sqrt{\frac{a}{b}} = q \quad \ldots \ldots \quad (3)$$

and since the second members are the same in Equations (1) and (3), the first members are equal; that is,

---

139. To what is the square root of the product of two quantities equal?
140. To what is the square root of the quotient of two quantities equal?

*The square root of the quotient of two quantities is equal to the quotient of their square roots.*

These principles enable us to *transform* radical expressions, or to *reduce* them to simpler forms; thus, the expression,

$$98ab^4 = 49b^4 \times 2a;$$

hence,  $\sqrt{98ab^4} = \sqrt{49b^4 \times 2a};$

and by the principle of (Art. 139),

$$\sqrt{49b^4 \times 2a} = \sqrt{49b^4} \times \sqrt{2a} = 7b^2\sqrt{2a}.$$

In like manner,

$$\sqrt{45a^2b^3c^2d} = \sqrt{9a^2b^2c^2 \times 5bd} = 3abc\sqrt{5bd}.$$

$$\sqrt{864a^2b^5c^{11}} = \sqrt{144a^2b^4c^{10} \times 6bc} = 12ab^2c^5\sqrt{6bc}.$$

The COEFFICIENT of a radical is the quantity without the sign ; thus, in the expressions,

$$7b^2\sqrt{2a}, \quad 3abc\sqrt{5bd}, \quad 12ab^2c^5\sqrt{6bc},$$

the quantities $7b^2$, $3abc$, $12ab^2c^5$, are *coefficients of the radicals*.

**141.** Hence, to simplify a radical of the second degree, we have the following

### RULE.

I. *Divide the expression under the radical sign into two factors, one of which shall be a perfect square :*

II. *Extract the square root of the perfect square, and then multiply this root by the indicated square root of the remaining factor.*

---

141. Give the rule for simplifying radicals of the second degree. How do you determine whether a given number has a factor which is a perfect square ?

Note.—To determine if a given number has any factor which is a perfect square, we examine and see if it is divisible by either of the perfect squares,

$$4, \quad 9, \quad 16, \quad 25, \quad 36, \quad 49, \quad 64, \quad 81, \quad \&c.;$$

if it is not, we conclude that it does not contain a factor which is a perfect square.

### EXAMPLES.

Reduce the following radicals to their simplest form:

1. $\sqrt{75a^3bc}$.          *Ans.* $5a\sqrt{3abc}$.

2. $\sqrt{128b^5a^6d^2}$.      *Ans.* $8b^2a^3d\sqrt{2b}$.

3. $\sqrt{32a^9b^8c}$.        *Ans.* $4a^4b^4\sqrt{2ac}$.

4. $\sqrt{256a^2b^4c^8}$.       *Ans.* $16ab^2c^4$.

5. $\sqrt{1024a^9b^7c^5}$.     *Ans.* $32a^4b^3c^2\sqrt{abc}$.

6. $\sqrt{729a^7b^5c^6d}$.     *Ans.* $27a^3b^2c^3\sqrt{abd}$.

7. $\sqrt{675a^7b^5c^2d}$.     *Ans.* $15a^3b^2c\sqrt{3abd}$.

8. $+\sqrt{1445a^3c^8d^4}$.     *Ans.* $17ac^4d^2\sqrt{5a}$.

9. $\sqrt{1008a^9d^7m^8}$.     *Ans.* $12a^4d^3m^4\sqrt{7ad}$.

10. $\sqrt{2156a^{10}b^8c^6}$.    *Ans.* $14a^5b^4c^3\sqrt{11}$.

11. $\sqrt{405a^7b^6d^8}$.     *Ans.* $9a^3b^3d^4\sqrt{5a}$.

**142.** Notes.—1. A *coefficient*, or a *factor of a coefficient*, may be carried under the radical sign, *by squaring it.* Thus,

1. $3a^2\sqrt{bc} = \sqrt{(3a^2)^2 \times bc} = \sqrt{9a^4bc}$.

2. $2ab\sqrt{d} = 2\sqrt{a^2b^2d} = \sqrt{4a^2b^2d}$.

---

142. How may a coefficient or factor be carried under the radical sign. To what is the square root of a negative quantity equal?

3. $4(a+b)\sqrt{a-b}=4\sqrt{(a+b)^2(a-b)}=4\sqrt{(a^2-b^2)(a+b)}$

4. $5bc\sqrt{a^2-c^2}=5\sqrt{b^2c^2(a^2-c^2)}$.

2. The square root of a negative quantity may also be simplified; thus,

$$\sqrt{-9}=\sqrt{9\times-1}=\sqrt{9}\times\sqrt{-1}=3\sqrt{-1},$$

and, $\sqrt{-4a^2}=\sqrt{4a^2}\times\sqrt{-1}=2a\sqrt{-1};$ also,

$$\sqrt{-8a^2b}=\sqrt{4a^2\times-2b}=2a\sqrt{-2b}=2a\sqrt{2b}\times\sqrt{-1};$$

that is, *the square root of a negative quantity is equal to the square root of the same quantity with a positive sign, multiplied into the square root of* $-1$.

Reduce the following:

1. $\sqrt{-64a^2b^2}$.  *Ans.* $8ab\sqrt{-1}$.

2. $\sqrt{-128a^4b^5}$.  *Ans.* $8a^2b^2\sqrt{2b}\sqrt{-1}$.

3. $\sqrt{-72a^5b^7c^6}$.  *Ans.* $6a^2b^3c^3\sqrt{2ab}\sqrt{-1}$.

4. $\sqrt{-48a^3bc^5}$.  *Ans.* $4ac^2\sqrt{3abc}\sqrt{-1}$.

---

### ADDITION OF RADICALS.

**143.** SIMILAR RADICALS, of the second degree, are those in which the quantities under the sign are the same. Thus, the radicals $3\sqrt{b}$, and $5c\sqrt{b}$ are similar, and so also are $9\sqrt{2}$, and $7\sqrt{2}$.

**144.** Radicals are added like other algebraic quantities; hence, the following

143. What are similar radicals of the second degree?
144. Give the rule for the addition of radicals of the second degree?

## RULE.

I. *If the radicals are similar, add their coefficients, and to the sum annex the common radical:*

II. *If the radicals are not similar, connect them together with their proper signs.*

Thus, $\quad 3a\sqrt{b} + 5c\sqrt{b} = (3a + 5c)\sqrt{b}.$

In like manner,

$$7\sqrt{2a} + 3\sqrt{2a} = (7 + 3)\sqrt{2a} = 10\sqrt{2a}.$$

NOTES.—1. Two radicals, which do not appear to be similar at first sight, may become so by transformation (Art. 141.)

For example,

$$\sqrt{48ab^2} + b\sqrt{75a} = 4b\sqrt{3a} + 5b\sqrt{3a} = 9b\sqrt{3a};$$

$$2\sqrt{45} + 3\sqrt{5} = 6\sqrt{5} + 3\sqrt{5} = 9\sqrt{5}.$$

2. When the radicals are not similar, the addition or subtraction can only be indicated. Thus, in order to add $3\sqrt{b}$ to $5\sqrt{a}$, we write,

$$5\sqrt{a} + 3\sqrt{b}.$$

Add together the following:

1. $\sqrt{27a^2}$ and $\sqrt{48a^2}$. $\hfill$ *Ans.* $7a\sqrt{3}.$

2. $\sqrt{50a^4b^2}$ and $\sqrt{72a^4b^2}$. $\hfill$ *Ans.* $11a^2b\sqrt{2}.$

3. $\sqrt{\dfrac{3a^2}{5}}$ and $\sqrt{\dfrac{a^2}{15}}.$ $\hfill$ *Ans.* $4a\sqrt{\dfrac{1}{15}}.$

4. $\sqrt{125}$ and $\sqrt{500a^2}$. $\hfill$ *Ans.* $(5 + 10a)\sqrt{5}.$

5. $\sqrt{\dfrac{50}{147}}$ and $\sqrt{\dfrac{100}{204}}$ $\hfill$ *Ans.* $\dfrac{10}{21}\sqrt{6}$

6. $\sqrt{98a^2x}$ and $\sqrt{36x^2 - 36a^2}$.

$Ans.$ $7a\sqrt{2x} + 6\sqrt{x^2 - a^2}$.

7. $\sqrt{98a^2x}$ and $\sqrt{288a^4x^5}$. $Ans.$ $(7a + 12a^2x^2)\sqrt{2x}$.

8. $\sqrt{72}$ and $\sqrt{128}$. $Ans.$ $14\sqrt{2}$.

9. $\sqrt{27}$ and $\sqrt{147}$. $Ans.$ $10\sqrt{3}$.

10. $\sqrt{\dfrac{2}{3}}$ and $\sqrt{\dfrac{27}{50}}$. $Ans.$ $\dfrac{19}{30}\sqrt{6}$.

11. $2\sqrt{a^2b}$ and $3\sqrt{64bx^4}$. $Ans.$ $(2a + 24x^2)\sqrt{b}$.

12. $\sqrt{243}$ and $10\sqrt{363}$. $Ans.$ $119\sqrt{3}$.

13. $\sqrt{320a^2b^2}$ and $\sqrt{245a^8b^6}$. $Ans.$ $(8ab + 7a^4b^3)\sqrt{5}$.

14. $\sqrt{75a^6b^7}$ and $\sqrt{300a^6b^5}$. $Ans.$ $(5a^3b^3 + 10a^3b^2)\sqrt{3b}$.

---

## SUBTRACTION OF RADICALS.

**145.** Radicals are subtracted like other algebraic quantities; hence, the following

### RULE.

I. *If the radicals are similar, subtract the coefficient of the subtrahend from that of the minuend, and to the difference annex the common radical:*

II. *If the radicals are not similar, indicate the operation by the minus sign.*

### EXAMPLES.

1. What is the difference between $3a\sqrt{b}$ and $a\sqrt{b}$?

Here, $3a\sqrt{b} - a\sqrt{b} = 2a\sqrt{b}$. $Ans.$

145. Give the rule for the subtraction of radicals.

2. From $9a\sqrt{27b^2}$ subtract $6a\sqrt{27b^2}$.

First, $9a\sqrt{27b^2} = 27ab\sqrt{3}$, and $6a\sqrt{27b^2} = 18ab\sqrt{3}$;

and, $27ab\sqrt{3} - 18ab\sqrt{3} = 9ab\sqrt{3}$. *Ans.*

Find the differences between the following:

3. $\sqrt{75}$ and $\sqrt{48}$.        *Ans.* $\sqrt{3}$.

4. $\sqrt{24a^2b^2}$ and $\sqrt{54b^4}$.     *Ans.* $(2ab - 3b^2)\sqrt{6}$.

5. $\sqrt{\dfrac{3}{5}}$ and $\sqrt{\dfrac{5}{27}}$.       *Ans.* $\dfrac{4}{45}\sqrt{15}$.

6. $\sqrt{128a^3b^2}$ and $\sqrt{32a^9}$.    *Ans.* $(8ab - 4a^4)\sqrt{2a}$.

7. $\sqrt{48a^3b^3}$ and $\sqrt{9ab}$.    *Ans.* $4ab\sqrt{3ab} - 3\sqrt{ab}$.

8. $\sqrt{242a^5b^5}$ and $\sqrt{2a^3b^3}$.    *Ans.* $(11a^2b^2 - ab)\sqrt{2ab}$.

9. $\sqrt{\dfrac{3}{4}}$ and $\sqrt{\dfrac{3}{9}}$.       *Ans.* $\dfrac{1}{6}\sqrt{3}$.

10. $\sqrt{320a^2}$ and $\sqrt{80a^2}$.      *Ans.* $4a\sqrt{5}$.

11. $\sqrt{720a^3b^3}$ and $\sqrt{245abc^2d^2}$.

                 *Ans.* $(12ab - 7cd)\sqrt{5ab}$.

12. $\sqrt{968a^2b^2}$ and $\sqrt{200a^2b^2}$.    *Ans.* $12ab\sqrt{2}$.

13. $\sqrt{112a^8b^6}$ and $\sqrt{28a^8b^6}$.    *Ans.* $2a^4b^3\sqrt{7}$.

---

## MULTIPLICATION OF RADICALS.

**146.** Radicals are multiplied like other algebraic quantities; hence, we have the following

### RULE.

I. *Multiply the coefficients together for a new coefficient:*

---

146. Give the rule for the multiplication of radicals.

II. *Multiply together the quantities under the radical signs:*

III. *Then reduce the result to its simplest form.*

1. Multiply $3a\sqrt{bc}$ by $2\sqrt{ab}$.

$$3a\sqrt{bc} \times 2\sqrt{ab} = 3a \times 2 \times \sqrt{bc} \times \sqrt{ab}.$$

which, by Art. 139, $\quad = 6a\sqrt{b^2ac} = 6ab\sqrt{ac}.$

Multiply the following:

2. $3\sqrt{5ab}$ and $4\sqrt{20a}$. 　　　　　 *Ans.* $120a\sqrt{b}$.

3. $2a\sqrt{bc}$ and $3a\sqrt{bc}$. 　　　　　 *Ans.* $6a^2bc$.

4. $2a\sqrt{a^2+b^2}$ and $-3a\sqrt{a^2+b^2}$. $A. -6a^2(a^2+b^2.)$

5. $2ab\sqrt{a+b}$ and $ac\sqrt{a-b}$. *Ans.* $2a^2bc\sqrt{a^2-b^2}$.

6. $3\sqrt{2}$ and $2\sqrt{8}$. 　　　　　　　 *Ans.* $24$.

7. $\frac{1}{3}\sqrt{\frac{1}{3}a^2b}$ and $\frac{2}{10}\sqrt{\frac{2}{3}c^2b}$. 　 *Ans.* $\frac{1}{30}abc\sqrt{15}$.

8. $2x + \sqrt{b}$ and $2x - \sqrt{b}$. 　 *Ans.* $4x^2 - b$.

9. $\sqrt{a + 2\sqrt{b}}$ and $\sqrt{a - 2\sqrt{b}}$. 　 *Ans.* $\sqrt{a^2 - 4b}$.

10. $3a\sqrt{27a^3}$ by $\sqrt{2a}$. 　　　　 *Ans.* $9a^3\sqrt{6}$.

---

## DIVISION OF RADICALS.

**147.** Radical quantities are divided like other algebraic quantities; hence, we have the following

### RULE.

I. *Divide the coefficient of the dividend by the coefficient of the divisor, for a new coefficient:*

---

147. Give the rule for the division of radicals.

II. *Divide the quantities under the radicals, in the same manner:*

III. *Then reduce the result to its simplest form.*

1. Divide $8a\sqrt{b^3c}$ by $4a\sqrt{bc^3}$.

$$\frac{8a}{4a} = 2, \quad \text{new coefficient.}$$

Art. 140, $\quad \dfrac{\sqrt{b^3c}}{\sqrt{bc^3}} = \sqrt{\dfrac{b^3c}{bc^3}} = \sqrt{\dfrac{b^2}{c^2}} = \dfrac{b}{c};$

hence, the quotient is $\quad 2 \times \dfrac{b}{c} = \dfrac{2b}{c}.$

2. Divide $5a\sqrt{b}$ by $2b\sqrt{c}$. $\qquad$ *Ans.* $\dfrac{5a}{2b}\sqrt{\dfrac{b}{c}}.$

3. Divide $12ac\sqrt{6bc}$ by $4c\sqrt{2b}$. $\qquad$ *Ans.* $3a\sqrt{3c}.$

4. Divide $6a\sqrt{96b^4}$ by $3\sqrt{8b^2}$. $\qquad$ *Ans.* $4ab\sqrt{3}.$

5. Divide $4a^2\sqrt{50b^5}$ by $2a^2\sqrt{5b}$. $\qquad$ *Ans.* $2b^2\sqrt{10}.$

6. Divide $26a^3b\sqrt{81a^2b^2}$ by $13a\sqrt{9ab}$. $\quad$ *A.* $6a^2b\sqrt{ab}.$

7. Divide $84a^3b^4\sqrt{27ac}$ by $42ab\sqrt{3a}$. $\quad$ *A.* $6a^2b^3\sqrt{c}.$

8. Divide $\sqrt{\tfrac{1}{4}a^2}$ by $\sqrt{2}$. $\qquad$ *Ans.* $\tfrac{1}{4}a.$

9. Divide $6a^2b^2\sqrt{20a^3}$ by $12\sqrt{5a}$. $\qquad$ *Ans.* $a^3b^2.$

10. Divide $6a\sqrt{10b^2}$ by $3\sqrt{5}$. $\qquad$ *Ans.* $2ab\sqrt{2}.$

11. Divide $48b^4\sqrt{15}$ by $2b^2\sqrt{\tfrac{1}{15}}$ $\qquad$ *Ans.* $360b^2.$

12. Divide $8a^2b^4c^3\sqrt{7d^3}$ by $2a\sqrt{28d}$. $\quad$ *Ans.* $2ab^4c^3d.$

13. Divide $96a^4c^3\sqrt{98b^5}$ by $48abc\sqrt{2b}$. $\quad$ *A.* $14a^3bc^2.$

**14.** Divide $27a^5b^6\sqrt{21a^3}$ by $\sqrt{7a}$.   *Ans.* $27a^6b^6\sqrt{3}$.

**15.** Divide $18a^8b^6\sqrt{8a^4}$ by $6ab\sqrt{a^2}$.   *Ans.* $6a^8b^5\sqrt{2}$.

## SQUARE ROOT OF POLYNOMIALS.

**148.** Before explaining the rule for the extraction of the square root of a polynomial, let us first examine the squares of several polynomials: we have,

$$(a + b)^2 = a^2 + 2ab + b^2,$$
$$(a + b + c)^2 = a^2 + 2ab + b^2 + 2(a + b)c + c^2,$$
$$(a + b + c + d)^2 = a^2 + 2ab + b^2 + 2(a + b)c + c^2$$
$$+ 2(a + b + c)d + d^2.$$

The *law* by which these squares are formed can be enunciated thus:

*The square of any polynomial is equal to the square of the first term, plus twice the product of the first term by the second, plus the square of the second; plus twice the first two terms multiplied by the third, plus the square of the third; plus twice the first three terms multiplied by the fourth, plus the square of the fourth; and so on.*

**149.** Hence, to extract the square root of a polynomial, we have the following

### RULE.

I. *Arrange the polynomial with reference to one of its letters, and extract the square root of the first term: this will give the first term of the root:*

---

148. What is the square of a binomial equal to? What is the square of a trinomial equal to? To what is the square of any polynomial equal?

149. Give the rule for extracting the square root of a polynomial? What is the first step? What the second? What the third? What the fourth?

ɔ

II. *Divide the second term of the polynomial by double the first term of the root, and the quotient will be the second term of the root :*

III. *Then form the square of the algebraic sum of the two terms of the root found, and subtract it from the first polynomial, and then divide the first term of the remainder by double the first term of the root, and the quotient will be the third term :*

IV. *Form the double product of the sum of the first and second terms by the third, and add the square of the third ; then subtract this result from the last remainder, and divide the first term of the result so obtained, by double the first term of the root, and the quotient will be the fourth term. Then proceed in a similar manner to find the other terms.*

### EXAMPLES.

1. Extract the square root of the polynomial,

$$49a^2b^2 - 24ab^3 + 25a^4 - 30a^3b + 16b^4.$$

First arrange it with reference to the letter $a$.

$$
\begin{array}{l|l}
25a^4 - 30a^3b + 49a^2b^2 - 24ab^3 + 16b^4 & 5a^2 - 3ab + 4b^2 \\
25a^4 - 30a^3b + \phantom{0}9a^2b^2 & 10a^2 \\
\hline
\phantom{25a^4 - 30a^3b + }40a^2b^2 - 24ab^3 + 16b^4 \quad . \quad . \quad \text{1st \textit{Rem.}} \\
\phantom{25a^4 - 30a^3b + }40a^2b^2 - 24ab^3 + 16b^4 \\
\hline
\phantom{25a^4 - 30a^3b + }0 \quad . \quad . \quad . \quad . \quad . \quad \text{2d \textit{Rem.}}
\end{array}
$$

After having arranged the polynomial with reference to $a$, extract the square root of $25a^4$; this gives $5a^2$, which is placed at the right of the polynomial : then divide the second term, $-30a^3b$, by the double of $5a^2$, or $10a^2$; the quotient is $-3ab$, which is placed at the right of $5a^2$. Hence, the first two terms of the root are $5a^2 - 3ab$. Squaring this binomial, it becomes $25a^4 - 30a^3b + 9a^2b^2$, which, subtracted from the proposed polynomial, gives a remainder, of which the first term is $40a^2b^2$. Dividing this first term by $10a^2$, (the double of $5a^2$), the quotient is $+ 4b^2$; this is the third term of the root, and is written on the right of the first two terms. By forming the double product of $5a^2 - 3ab$ by $4b^2$, squaring $4b^2$, and taking the sum, we find the polynomial $40a^2b^2 - 24ab^3 + 16b^4$, which, subtracted from the first remainder, gives 0. Therefore, $5a^2 - 3ab + 4b^2$ is the required root.

2. Find the square root of $a^4 + 4a^3x + 6a^2x^2 + 4ax^3 + x^4$.

$Ans.$ $a^2 + 2ax + x^2$.

3. Find the square root of $a^4 - 4a^3x + 6a^2x^2 - 4ax^3 + x^4$.

$Ans.$ $a^2 - 2ax + x^2$.

4. Find the square root of

$$4x^6 + 12x^5 + 5x^4 - 2x^3 + 7x^2 - 2x + 1.$$

$Ans.$ $2x^3 + 3x^2 - x + 1$.

5. Find the square root of

$$9a^4 - 12a^3b + 28a^2b^2 - 16ab^3 + 16b^4.$$

$Ans.$ $3a^2 - 2ab + 4b^2$.

6. What is the square root of

$$x^4 - 4ax^3 + 4a^2x^2 - 4x^2 + 8ax + 4 ?$$

$Ans.$ $x^2 - 2ax - 2$.

7. What is the square root of

$$9x^2 - 12x + 6xy + y^2 - 4y + 4 ?$$

$Ans.$ $3x + y - 2$.

8. What is the square root of $y^4 - 2y^2x^2 + 2x^2 - 2y^2 + 1 + x^4 ?$ $Ans.$ $y^2 - x^2 - 1$.

9. What is the square root of $9a^4b^4 - 30a^3b^3 + 25a^2b^2 ?$

$Ans.$ $3a^2b^2 - 5ab$.

10. Find the square root of

$$25a^4b^2 - 40a^3b^2c + 76a^2b^2c^2 - 48ab^2c^3 + 36b^2c^4 - 30a^4bc$$
$$+ 24a^3bc^2 - 36a^2bc^3 + 9a^4c^2.$$

$Ans.$ $5a^2b - 3a^2c - 4abc + 6bc^2$.

**150.** We will conclude this subject with the following remarks:

1st. A binomial can never be a perfect square, since we know that the square of the most simple polynomial, viz., a binomial, contains three distinct parts, which cannot experience any reduction amongst themselves. Thus, the expression $a^2 + b^2$, is not a perfect square; it wants the term $\pm 2ab$, in order that it should be the square of $a \pm b$.

2d. In order that a trinomial, when arranged, may be a perfect square, its two extreme terms must be squares, and the middle term must be the double product of the square roots of the two others. Therefore, to obtain the square root of a trinomial when it is a perfect square: *Extract the roots of the two extreme terms, and give these roots the same or contrary signs, according as the middle term is positive or negative. To verify it, see if the double product of the two roots is the same as the middle term of the trinomial.* Thus,

$$9a^6 - 48a^4b^2 + 64a^2b^4, \text{ is a perfect square,}$$

since, $\sqrt{9a^6} = 3a^3$, and $\sqrt{64a^2b^4} = -8ab^2$;

and also,

$$2 \times 3a^3 \times -8ab^2 = -48a^4b^2 = \text{ the middle term.}$$

But, $4a^2 + 14ab + 9b^2$ is not a perfect square: for, although $4a^2$ and $+9b^2$ are the squares of $2a$ and $3b$, yet $2 \times 2a \times 3b$ is not equal to $14ab$.

3d. In the series of operations required by the general rule, when the first term of one of the remainders is not exactly divisible by twice the first term of the root, we may

---

150. Can a binomial ever be a perfect power? Why not? When is a trinomial a perfect square? When, in extracting the square root, we find that the first term of the remainder is not divisible by twice the root, is the polynomial a perfect power or not?

conclude that the proposed polynomial is not a perfect square. This is an evident consequence of the course of reasoning by which we have arrived at the general rule for extracting the square root.

4th. When the polynomial is *not a perfect square*, it may sometimes be simplified (See Art. 139).

Take, for example, the expression, $\sqrt{a^3b + 4a^2b^2 + 4ab^3}$.

The quantity under the radical is not a perfect square; but it can be put under the form $ab(a^2 + 4ab + 4b^2.)$ Now, the factor within the parenthesis is evidently the square of $a + 2b$, whence, we may conclude that,

$$\sqrt{a^3b + 4a^2b^2 + 4ab^3} = (a + 2b)\sqrt{ab}.$$

2. Reduce $\sqrt{2a^2b - 4ab^2 + 2b^3}$ to its simplest form.

$$Ans. \ (a - b)\sqrt{2b}.$$

# CHAPTER VIII.

## EQUATIONS OF THE SECOND DEGREE.

### EQUATIONS CONTAINING ONE UNKNOWN QUANTITY.

**151.** AN EQUATION of the second degree containing bu. one unknown quantity, is one in which the greatest exponent is equal to 2. Thus,

$$x^2 = a, \qquad ax^2 + bx = c,$$

are equations of the second degree.

**152.** Let us see to what form every equation of the second degree may be reduced.

Take any equation of the second degree, as,

$$(1 + x)^2 - \frac{3}{4}x - 10 = 5 - \frac{x}{4} + \frac{x^2}{2}.$$

Clearing of fractions, and performing indicated operations, we have,

$$4 + 8x + 4x^2 - 3x - 40 = 20 - x + 2x^2.$$

Transposing the unknown terms to the first member, the known terms to the second, and arranging with reference to the powers of $x$, we have,

$$4x^2 - 2x^2 + 8x - 3x + x = 20 + 40 - 4;$$

151. What is an equation of the second degree ?  Give an example.
152. To what form may every equation of the second degree be reduced?

and, by reducing,

$$2x^2 + 6x = 56;$$

dividing by the coefficient of $x^2$, we have,

$$x^2 + 3x = 28.$$

If we denote the coefficient of $x$ by $2p$, and the second member by $q$, we have,

$$x^2 + 2px = q.$$

This is called the *reduced equation*.

**153.** When the reduced equation is of this form, it contains three terms, and is called a *complete equation*. The terms are,

FIRST TERM.—The second power of the unknown quantity, with a plus sign.

SECOND TERM.—The first power of the unknown quantity, with a coefficient.

THIRD TERM.—A known term, in the second member.

Every equation of the second degree may be reduced to this form, by the following

### RULE.

I. *Clear the equation of fractions, and perform all the indicated operations:*

II. *Transpose all the unknown terms to the first member, and all the known terms to the second member:*

---

153. How many terms are there in a complete equation? What is the first term? What is the second term? What is the third term? How many operations are there in reducing an equation of the second degree to the required form? What is the first? What the second? What the third? What the fourth?

III. *Reduce all the terms containing the square of the unknown quantity to a single term, one factor of which is the square of the unknown quantity; reduce, also, all the terms containing the first power of the unknown quantity, to a single term:*

IV. *Divide both members of the resulting equation by the coefficient of the square of the unknown quantity.*

**154.** A Root of an equation is such a value of the unknown quantity as, being substituted for it, will satisfy the equation; that is, make the two members equal.

The Solution of an equation is the operation of finding its roots.

### INCOMPLETE EQUATIONS.

**155.** It may happen, that $2p$, the coefficient of the first power of $x$, in the equation $x + 2px = q$, is equal to 0. In this case, the first power of $x$ will disappear, and the equation will take the form,

$$x^2 = q \quad \cdots \quad \cdots \quad (1.)$$

This is called an incomplete equation; hence,

An incomplete equation, when reduced, contains but two terms; the square of the unknown quantity, and a known term.

**156.** Extracting the square root of both members of Equation (1), we have,

$$x = \pm \sqrt{q}.$$

154. What is the root of an equation? What is the solution of an equation?

155. What form will the reduced equation take when the coefficient of $x$ is 0? What is the equation then called? How many terms are there in an incomplete equation? What are they?

156. What is the rule for the solution of an incomplete equation? How many roots are there in every incomplete equation? How do the roots compare with each other?

Hence, for the solution of incomplete equations:

I. *Reduce the equation to the form* $x^2 = q$:

II. *Then extract the square root of both members.*

NOTE.—There will be two roots, numerically equal, but having contrary signs. Denoting the first by $x'$, and the second by $x''$, we have,

$$x' = +\sqrt{q}, \quad \text{and} \quad x'' = -\sqrt{q}.$$

Substituting $+\sqrt{q}$, or $-\sqrt{q}$, for $x$, in Equation (1), we have,

$$(+\sqrt{q})^2 = q; \quad \text{and,} \quad (-\sqrt{q})^2 = q;$$

hence, both satisfy the equation; they are, therefore, roots. (Art. 154.)

1. What are the values of $x$ in the equation,

$$3x^2 + 8 = 5x^2 - 10?$$

By transposing, $3x^2 - 5x^2 = -10 - 8.$

Reducing, $\qquad -2x^2 = -18.$

Dividing by $-2$, $\qquad x^2 = 9.$

Extracting square root, $x = \pm\sqrt{9} = +3$ and $-3.$

Hence, $\qquad x' = +3,$ and $x'' = -3.$

2. What are the roots of the equation,

$$3x^2 + 6 = 4x^2 - 10?$$

$$\textit{Ans. } x' = +4, \quad x'' = -4.$$

3. What are the roots of the equation,

$$\frac{1}{3}x^2 - 8 = \frac{x^2}{9} + 10 ?$$

*Ans.* $x' = +9,\ x'' = -9.$

4. What are the roots of the equation,

$$4x^2 + 13 - 2x^2 = 45 ?$$

*Ans.* $x' = +4,\ x'' = -4.$

5. What are the roots of the equation,

$$6x^2 - 7 = 3x^2 + 5 ?$$

*Ans.* $x' = +2,\ x'' = -2.$

6. What are the roots of the equation,

$$8 + 5x^2 = \frac{x^2}{5} + 4x^2 + 28 ?$$

*Ans.* $x' = +5,\ x'' = -5.$

7. What are the roots of the equation,

$$\frac{3x^2 + 5}{8} - \frac{x^2 + 29}{3} = 117 - 5x^2 ?$$

*Ans.* $x' = +5,\ x'' = -5.$

8. What are the roots of the equation,

$$x^2 + ab = 5x^2 ?$$

*Ans.* $x' = +\frac{1}{2}\sqrt{ab},\ x'' = -\frac{1}{2}\sqrt{ab}.$

9. What are the roots of the equation,

$$x\sqrt{a + x^2} = b + x^2 ?$$

*Ans.* $x' = \dfrac{b}{\sqrt{a - 2b}},\ x'' = -\dfrac{b}{\sqrt{a - 2b}}$

### PROBLEMS.

1. What number is that which being multiplied by itself the product will be 144?

Let $x =$ the number: then,

$$x \times x = x^2 = 144.$$

It is plain that the value of $x$ will be found by extracting the square root of both members of the equation: that is,

$$\sqrt{x^2} = \sqrt{144} : \text{that is, } x = 12.$$

2. A person being asked how much money he had, said, if the number of dollars be squared and 6 be added, the sum will be 42: how much had he?

Let $x =$ the number of dollars.

Then, by the conditions,

$$x^2 + 6 = 42;$$

hence, $\qquad x^2 = 42 - 6 = 36,$

and, $\qquad\qquad x = 6. \qquad Ans. \text{ \$6.}$

3. A grocer being asked how much sugar he had sold to a person, answered, if the square of the number of pounds be multiplied by 7, the product will be 1575. How many pounds had he sold?

Denote the number of pounds by $x$. Then, by the conditions of the question,

$$7x^2 = 1575;$$

hence, $\qquad x^2 = 225,$

and, $\qquad\qquad x = 15. \qquad Ans. \text{ 15.}$

4. A person being asked his age, said, if from the square

of my age in years, you take 192 years, the remainder will be the square of half my age: what was his age?

Denote the number of years in his age by $x$.

Then, by the conditions of the question,

$$x^2 - 192 = \left(\frac{1}{2}x\right)^2 = \frac{x^2}{4},$$

and by clearing the fractions,

$$4x^2 - 768 = x^2;$$

hence, $\qquad 4x^2 - x^2 = 768,$

and, $\qquad\qquad 3x^2 = 768,$

$$x^2 = 256,$$

$$x = 16. \qquad Ans.\ 16\ \text{years.}$$

5. What number is that whose eighth part multiplied by its fifth part and the product divided by 4, will give a quotient equal to 40?

Let $x =$ the number.

By the conditions of the question,

$$\left(\frac{1}{8}x \times \frac{1}{5}x\right) \div 4 = 40;$$

hence, $\qquad\qquad \dfrac{x^2}{160} = 40;$

by clearing of fractions,

$$x^2 = 6400,$$

$$x = 80. \qquad Ans.\ 80.$$

6. Find a number such that one-third of it multiplied by one fourth shall be equal to 108. $\qquad Ans.\ 36.$

7. What number is that whose sixth part multiplied by its fifth part and the product divided by ten, will give a quotient equal to 3? $\qquad Ans.\ 30.$

8. What number is that whose square, plus 18, will be equal to half the square, plus $30\frac{1}{2}$ ? *Ans.* 5.

9. What numbers are those which are to each other as 1 to 2, and the difference of whose squares is equal to 75 ?

Let $x$ = the less number.

Then, $2x$ = the greater.

Then, by the conditions of the question,

$$4x^2 - x^2 = 75 ;$$

hence, $$3x^2 = 75,$$

and by dividing by 3, $x^2 = 25$, and $x = 5$,

and, $$2x = 10.$$

*Ans.* 5 and 10.

10. What two numbers are those which are to each other as 5 to 6, and the difference of whose squares is 44 ?

Let $x$ = the greater number.

Then, $\frac{5}{6}x$ = the less.

By the conditions of the problem,

$$x^2 - \frac{25}{36}x^2 = 44 ;$$

by clearing of fractions,

$$36x^2 - 25x^2 = 1584 ;$$

hence, $$11x^2 = 1584,$$

and, $$x^2 = 144 ;$$

hence, $$x = 12,$$

and, $$\frac{5}{6}x = 10.$$

*Ans.* 10 and 12.

11. What two numbers are those which are to each other as 3 to 4, and the difference of whose squares is 28 ?

*Ans.* 6 and 8.

12. What two numbers are those which are to each other as 5 to 11, and the sum of whose squares is 584 ?

*Ans.* 10 and 22.

13. *A* says to *B*, my son's age is one quarter of yours, and the difference between the squares of the numbers representing their ages is 240 : what were their ages ?

*Ans.* $\begin{cases} \text{Eldest,} & 16, \\ \text{Younger,} & 4. \end{cases}$

### *Two unknown quantities.*

**157.** When there are two or more unknown quantities:

I. *Eliminate one of the unknown quantities by* Art. 113 :

II. *Then extract the square root of both members of the equation.*

### PROBLEMS.

1. There is a room of such dimensions, that the difference of the sides multiplied by the less, is equal to 36, and the product of the sides is equal to 360 : what are the sides ?

Let $x =$ the length of the less side ;

$y =$ the length of the greater.

Then, by the first condition,

$$(y - x)x = 36 ;$$

and by the 2d,    $$xy = 360.$$

---

157. How do you proceed when there are two or more unknown quantities?

From the first equation, we have,

$$xy - x^2 = 36;$$

and by subtraction, $\qquad x^2 = 324.$

Hence, $\qquad x = \sqrt{324} = 18;$

$$y = \frac{360}{18} = 20.$$

*Ans.* $x = 18, \; y = 20.$

2. A merchant sells two pieces of muslin, which together measure 12 yards. He received for each piece just so many dollars per yard as the piece contained yards. Now, he gets four times as much for one piece as for the other: how many yards in each piece?

Let $x =$ the number of yards in the larger piece;

$\qquad y =$ the number of yards in the shorter piece.

Then, by the conditions of the question,

$$x + y = 12.$$

$x \times x = x^2 =$ what he got for the larger piece;

$y \times y = y^2 =$ what he got for the shorter;

and, $\qquad x^2 = 4y^2,$ by the 2d condition,

$\qquad x = 2y,$ by extracting the square root.

Substituting this value of $x$ in the first equation, we have,

$$y + 2y = 12;$$

and, consequently, $\qquad y = 4,$

and, $\qquad x = 8.$

*Ans.* 8 and 4.

3. What two numbers are those whose product is 30, and the quotient of the greater by the less, $3\frac{1}{3}$? *Ans.* 10 and 3.

4. The product of two numbers is $a$, and their quotient $b$: what are the numbers? 

*Ans.* $\sqrt{ab}$, and $\sqrt{\dfrac{a}{b}}$.

5. The sum of the squares of two numbers is 117, and the difference of their squares 45 : what are the numbers?

*Ans.* 9 and 6.

6. The sum of the squares of two numbers is $a$, and the difference of their squares is $b$: what are the numbers?

$$Ans. \quad x = \sqrt{\frac{a+b}{2}}, \quad y = \sqrt{\frac{a-b}{2}}.$$

7. What two numbers are those which are to each other as 3 to 4, and the sum of whose squares is 225 ?

*Ans.* 9 and 12.

8. What two numbers are those which are to each other as $m$ to $n$, and the sum of whose squares is equal to $a^2$ ?

$$Ans. \quad \frac{ma}{\sqrt{m^2+n^2}}, \quad \frac{na}{\sqrt{m^2+n^2}}$$

9. What two numbers are those which are to each other as 1 to 2, and the difference of whose squares is 75 ?

*Ans.* 5 and 10.

10. What two numbers are those which are to each other as $m$ to $n$, and the difference of whose squares is equal to $b^2$ ?

$$Ans. \quad \frac{mb}{\sqrt{m^2-n^2}}, \quad \frac{nb}{\sqrt{m^2-n^2}}.$$

11. A certain sum of money is placed at interest for six months, at 8 per cent. per annum. Now, if the sum put at interest be multiplied by the number expressing the interest, the product will be $562500 : what is the principal at interest?      *Ans.* $3750.

12. A person distributes a sum of money between a number of women and boys. The number of women is to the number of boys as 3 to 4. Now, the boys receive one-half as many dollars as there are persons, and the women, twice as many dollars as there are boys, and together they receive

138 dollars: how many women were there, and how many boys?

$$Ans. \begin{cases} 36 \text{ women.} \\ 48 \text{ boys.} \end{cases}$$

## COMPLETE EQUATIONS.

**158.** The reduced form of the complete equation (Art. 153) is,

$$x^2 + 2px = q.$$

Comparing the first member of this equation with the square of a binomial (Art. 54), we see that it needs but the square of half the coefficient of $x$, to make it a perfect square. Adding $p^2$ to both members (Ax. 1, Art. 102), we have,

$$x^2 + 2px + p^2 = q + p^2.$$

Then, extracting the square root of both members (Ax. 5), we have,

$$x + p = \pm\sqrt{q + p^2}.$$

Transposing $p$ to the second member, we have,

$$x = -p \pm \sqrt{q + p^2}.$$

Hence, there are two roots, one corresponding to the *plus* sign of the radical, and the other to the *minus* sign. Denoting these roots by $x'$ and $x''$, we have,

$$x' = -p + \sqrt{q + p^2}, \text{ and } x'' = -p - \sqrt{q + p^2}.$$

The root denoted by $x'$ is called the *first* root; that denoted by $x''$ is called the *second root*.

---

158. What is the form of the reduced equation of the second degree? What is the square of the binomial $x + p$? How many of those terms are found in the first term of the reduced equation? What must be added to make the first member a perfect square? How many roots are there in every equation of the first degree? What is the first root equal to? What is the second equal to?

**159.** The operation of squaring half the coefficient of $x$ and adding the result to both members of the equation, is called *Completing the Square.* For the solution of every complete equation of the second degree, we have the following

### RULE.

I. *Reduce the equation to the form,* $x^2 + 2px = q$ :

II. *Take half the coefficient of the second term, square it, and add the result to both members of the equation :*

III. *Then extract the square root of both members ; after which, transpose the known term to the second member.*

NOTE.—Although, in the beginning, the student should complete the square and then extract the square root, yet he should be able, in all cases, to write the roots immediately, by the following   (See Art. 158)

### RULE.

I. *The first root is equal to half the coefficient of the second term of the reduced equation, taken with a contrary sign, plus the square root of the second member increased by the square of half the coefficient of the second term :*

II. *The second root is equal to half the coefficient of the second term of the reduced equation, taken with a contrary sign, minus the square root of the second member increased by the square of half the coefficient of the second term.*

**160.** We will now show that the complete equation of

---

159. What is the operation of completing the square ? How many operations are there in the solution of every equation of the second degree ? What is the first ? What the second ? What the third ? Give the rule for writing the roots without completing the square ?

160. How many forms will the complete equation of the second degree assume ? On what will these forms depend ? What are the signs of $2p$

the second degree will take four forms, dependent on the signs of $2p$ and $q$.

1st. Let us suppose $2p$ to be positive, and $q$ positive; we shall then have,

$$x^2 + 2px = q. \quad \ldots \ldots (1.)$$

2d. Let us suppose $2p$ to be negative, and $q$ positive; we shall then have,

$$x^2 - 2px = q. \quad \ldots \ldots (2.)$$

3d. Let us suppose $2p$ to be positive, and $q$ negative; we shall then have,

$$x^2 + 2px = -q. \quad \ldots \ldots (3.)$$

4th. Let us suppose $2p$ to be negative, and $q$ negative; we shall then have,

$$x^2 - 2px = -q. \quad \ldots \ldots (4.)$$

As these are all the combinations of signs that can take place between $2p$ and $q$, we conclude that every complete equation of the second degree will be reduced to one or the other of these four forms:

$$x^2 + 2px = +q, \quad \ldots \text{ 1st form.}$$
$$x^2 - 2px = +q, \quad \ldots \text{ 2d form.}$$
$$x^2 + 2px = -q, \quad \ldots \text{ 3d form.}$$
$$x^2 - 2px = -q, \quad \ldots \text{ 4th form.}$$

### EXAMPLES OF THE FIRST FORM.

1. What are the values of $x$ in the equation,

$$2x^2 + 8x = 64 ?$$

If we first divide by the coefficient 2, we obtain

$$x^2 + 4x = 32.$$

---

and $q$ in the first form? What in the second? What in the third? What in the fourth?

Then, completing the square,

$$x^2 + 4x + 4 = 32 + 4 = 36.$$

Extracting the root,

$$x + 2 = \pm\sqrt{36} = +6, \text{ and } -6.$$

Hence, $\quad\quad x' = -2 + 6 = +4;$

and, $\quad\quad x'' = -2 - 6 = -8.$

Hence, in this form, the smaller root, numerically, is positive, and the larger negative.

VERIFICATION.

If we take the positive value, viz.: $x' = +4$,

the equation, $\quad\quad x^2 + 4x = 32,$

gives $\quad\quad 4^2 + 4 \times 4 = 32;$

and if we take the negative value of $x$, viz.: $x'' = -8$,

the equation, $\quad\quad x^2 + 4x = 32,$

gives $\quad\quad (-8)^2 + 4(-8) = 64 - 32 = 32;$

from which we see that either of the values of $x$, viz.: $x' = +4$, or $x'' = -8$, will satisfy the equation.

2. What are the values of $x$ in the equation,

$$3x^2 + 12x - 19 = -x^2 - 12x + 89?$$

By transposing the terms, we have,

$$3x^2 + x^2 + 12x + 12x = 89 + 19;$$

and by reducing,

$$4x^2 + 24x = 108;$$

and dividing by the coefficient of $x^2$,

$$x^2 + 6x = 27.$$

Now, by completing the square,

$$x^2 + 6x + 9 = 36;$$

extracting the square root,

$$x + 3 = \pm\sqrt{36} = +6, \text{ and } -6;$$

hence, $\quad\quad x' = +6 - 3 = +3;$

and, $\quad\quad x'' = -6 - 3 = -9.$

VERIFICATION.

If we take the plus root, the equation,

$$x^2 + 6x = 27,$$

gives $\quad\quad (3)^2 + 6(3) = 27;$

and for the negative root,

$$x^2 + 6x = 27,$$

gives $\quad (-9)^2 + 6(-9) = 81 - 54 = 27.$

3. What are the values of $x$ in the equation,

$$x^2 - 10x + 15 = \frac{x^2}{5} - 34x + 155?$$

By clearing of fractions, we have,

$$5x^2 - 50x + 75 = x^2 - 170x + 775;$$

by transposing and reducing, we obtain,

$$4x^2 + 120x = 700;$$

then, dividing by the coefficient of $x^2$, we have,

$$x^2 + 30x = 175;$$

and by completing the square,

$$x^2 + 30x + 225 = 400;$$

and by extracting the square root,

$$x + 15 = \pm\sqrt{400} = +20, \text{ and } -20.$$

Hence, $\qquad x' = +5, \text{ and } x'' = -35.$

### VERIFICATION.

For the plus value of $x$, the equation,

$$x^2 + 30x = 175,$$

gives, $\qquad (5)^2 + 30 \times 5 = 25 + 150 = 175.$

And for the negative value of $x$, we have,

$$(-35)^2 + 30(-35) = 1225 - 1050 = 175.$$

4. What are the values of $x$ in the equation,

$$\frac{5}{6}x^2 - \frac{1}{2}x + \frac{3}{4} = 8 - \frac{2}{3}x - x^2 + \frac{273}{12}?$$

Clearing of fractions, we have,

$$10x^2 - 6x + 9 = 96 - 8x - 12x^2 + 273;$$

transposing and reducing,

$$22x^2 + 2x = 360;$$

dividing both members by 22,

$$x^2 + \frac{2}{22}x = \frac{360}{22}.$$

Add $\left(\frac{1}{22}\right)^2$ to both members, and the equation becomes,

$$x^2 + \frac{2}{22}x + \left(\frac{1}{22}\right)^2 = \frac{360}{22} + \left(\frac{1}{22}\right)^2;$$

whence, by extracting the square root,

$$x + \frac{1}{22} = \pm \sqrt{\frac{360}{22} + \left(\frac{1}{22}\right)^2},$$

therefore,

$$x' = -\frac{1}{22} + \sqrt{\frac{360}{22} + \left(\frac{1}{22}\right)^2},$$

and,

$$x'' = -\frac{1}{22} - \sqrt{\frac{360}{22} + \left(\frac{1}{22}\right)^2}.$$

It remains to perform the numerical operations. In the first place,

$$\frac{360}{22} + \left(\frac{1}{22}\right)^2,$$

must be reduced to a single number, having $(22)^2$ for its denominator. Now,

$$\frac{360}{22} + \left(\frac{1}{22}\right)^2 = \frac{360 \times 22 + 1}{(22)^2} = \frac{7921}{(22)^2};$$

extracting the square root of 7921, we find it to be 89; therefore,

$$\pm\sqrt{\frac{360}{22} + \left(\frac{1}{22}\right)^2} = \pm\frac{89}{22}.$$

Consequently, the plus value of $x$ is,

$$x' = -\frac{1}{22} + \frac{89}{22} = \frac{88}{22} = 4,$$

and the negative value is,

$$x'' = -\frac{1}{22} - \frac{89}{22} = -\frac{45}{11};$$

that is, one of the two values of $x$ which will satisfy the proposed equation is a positive whole number, and the other a negative fraction.

NOTE.—Let the pupil be exercised in writing the roots, in the last five, and in the following examples, *without completing the square.*

**5.** What are the values of $x$ in the equation,

$$3x^2 + 2x - 9 = 76 ?$$

$$Ans. \begin{cases} x' = 5. \\ x'' = - 5\tfrac{2}{3}. \end{cases}$$

**6.** What are the values of $x$ in the equation,

$$2x^2 + 8x + 7 = \frac{5x}{4} - \frac{x^2}{8} + 197 ?$$

$$Ans. \begin{cases} x' = 8. \\ x'' = - 11\tfrac{3}{17}. \end{cases}$$

**7.** What are the values of $x$ in the equation,

$$\frac{x^2}{4} - \frac{x}{3} + 15 = \frac{x^2}{9} - 8x + 95\tfrac{1}{4} ?$$

$$Ans. \begin{cases} x' = 9. \\ x'' = - 64\tfrac{1}{2}. \end{cases}$$

**8.** What are the values of $x$ in the equation,

$$\frac{x^2}{1} - \frac{5x}{4} - 8 = \frac{x}{2} - 7x + 6\tfrac{1}{2} ?$$

$$Ans. \begin{cases} x' = 2. \\ x'' = - 7\tfrac{1}{4}. \end{cases}$$

**9.** What are the values of $x$ in the equation,

$$\frac{x^2}{2} + \frac{x}{4} = \frac{x^2}{5} - \frac{x}{10} + \frac{13}{20}$$

$$Ans. \begin{cases} x' = 1. \\ x'' = - 2\tfrac{1}{4}. \end{cases}$$

### EXAMPLES OF THE SECOND FORM.

**1.** What are the values of $x$ in the equation,

$$x^2 - 8x + 10 = 19 ?$$

By transposing,

$$x^2 - 8x = 19 - 10 = 9;$$

then, by completing the square,

$$x^2 - 8x + 16 = 9 + 16 = 25;$$

and by extracting the root,

$$x - 4 = \pm \sqrt{25} = +5, \text{ or } -5.$$

Hence,

$$x' = 4 + 5 = 9, \text{ and } x'' = 4 - 5 = -1.$$

That is, in this form, the larger root, numerically, is positive, and the lesser negative.

### VERIFICATION.

If we take the positive value of $x$, the equation,

$x^2 - 8x = 9$, gives $(9)^2 - 8 \times 9 = 81 - 72 = 9;$

and if we take the negative value, the equation,

$x^2 - 8x = 9$, gives $(-1)^2 - 8(-1) = 1 + 8 = 9;$

from which we see that both roots alike satisfy the equation.

2. What are the values of $x$ in the equation,

$$\frac{x^2}{2} + \frac{x}{3} - 15 = \frac{x^2}{4} + x - 14\tfrac{3}{4}?$$

By clearing of fractions, we have,

$$6x^2 + 4x - 180 = 3x^2 + 12x - 177,$$

and by transposing and reducing,

$$3x^2 - 8x = 3;$$

and dividing by the coefficient of $x^2$, we obtain,

$$x^2 - \frac{8}{3}x = 1.$$

10

Then, by completing the square, we have,

$$x^2 - \frac{8}{3}x + \frac{16}{9} = 1 + \frac{16}{9} = \frac{25}{9};$$

and by extracting the square root,

$$x - \frac{4}{3} = \pm\sqrt{\frac{25}{9}} = +\frac{5}{3}, \text{ and } -\frac{5}{3}.$$

Hence,

$$x' = \frac{4}{3} + \frac{5}{3} = +3, \text{ and } x'' = \frac{4}{3} - \frac{5}{3} = -\frac{1}{3}.$$

### VERIFICATION.

For the positive root of $x$, the equation,

$$x^2 - \frac{8}{3}x = 1,$$

gives

$$3^2 - \frac{8}{3} \times 3 = 9 - 8 = 1;$$

and for the negative root, the equation,

$$x^2 - \frac{8}{3}x = 1,$$

gives

$$\left(-\frac{1}{3}\right)^2 - \frac{8}{3} \times -\frac{1}{3} = \frac{1}{9} + \frac{8}{9} = 1.$$

3. What are the values of $x$ in the equation,

$$\frac{x^2}{2} - \frac{x}{3} + 7\frac{2}{3} = 8?$$

Clearing of fractions, and dividing by the coefficient of $x^2$, we have,

$$x^2 - \frac{2}{3}x = 1\frac{1}{4}.$$

Completing the square, we have,

$$x^2 - \frac{2}{3}x + \frac{1}{9} = 1\tfrac{1}{4} + \frac{1}{9} = \frac{49}{36};$$

then, by extracting the square root, we have,

$$x - \frac{1}{3} = \pm\sqrt{\frac{49}{36}} = +\frac{7}{6}, \text{ and } -\frac{7}{6};$$

hence,

$$x' = \frac{1}{3} + \frac{7}{6} = \frac{9}{6} = 1\tfrac{1}{2}, \text{ and } x'' = \frac{1}{3} - \frac{7}{6} = -\frac{5}{6}.$$

VERIFICATION.

If we take the positive root of $x$, the equation,

$$x^2 - \frac{2}{3}x = 1\tfrac{1}{4},$$

gives $\quad (1\tfrac{1}{2})^2 - \frac{2}{3} \times 1\tfrac{1}{2} = 2\tfrac{1}{4} - 1 = 1\tfrac{1}{4};$

and for the negative root, the equation,

$$x^2 - \frac{2}{3}x = 1\tfrac{1}{4},$$

gives $\quad \left(-\frac{5}{6}\right)^2 - \frac{2}{3} \times -\frac{5}{6} = \frac{25}{36} + \frac{10}{18} = \frac{45}{36} = 1\tfrac{1}{4}$

4. What are the values of $x$ in the equation,

$$4a^2 - 2x^2 + 2ax = 18ab - 18b^2?$$

By transposing, changing the signs, and dividing by 2, the equation becomes,

$$x^2 - ax = 2a^2 - 9ab + 9b^2;$$

whence, completing the square,

$$x^2 - ax + \frac{a^2}{4} = \frac{9a^2}{4} - 9ab + 9b^2;$$

extracting the square root,

$$x = \frac{a}{2} \pm \sqrt{\frac{9a^2}{4} - 9ab + 9b^2}.$$

Now, the square root of $\frac{9a^2}{4} - 9ab + 9b^2$, is evidently $\frac{3a}{2} - 3b$. Therefore,

$$x = \frac{a}{2} \pm \left(\frac{3a}{2} - 3b\right), \text{ and } \begin{cases} x' = & 2a - 3b. \\ x'' = & -a + 3b. \end{cases}$$

What will be the numerical values of $x$, if we suppose $a = 6$, and $b = 1$?

5. What are the values of $x$ in the equation,

$$\frac{1}{3}x - 4 - x^2 + 2x - \frac{4}{5}x^2 = 45 - 3x^2 + 4x?$$

$$Ans. \begin{cases} x' = & 7.12 \\ x'' = & -5.73 \end{cases} \begin{matrix} \text{to within} \\ 0.01. \end{matrix}$$

6. What are the values of $x$ in the equation,

$$8x^2 - 14x + 10 = 2x + 34?$$

$$Ans. \begin{cases} x' = 3. \\ x'' = -1. \end{cases}$$

7. What are the values of $x$ in the equation,

$$\frac{x^2}{4} - 30 + x = 2x - 22?$$

$$Ans. \begin{cases} x' = 8. \\ x'' = -4. \end{cases}$$

8. What are the values of $x$ in the equation,

$$x^2 - 3x + \frac{x^2}{2} = 9x + 13\tfrac{1}{2}?$$

$$Ans. \begin{cases} x' = 9. \\ x'' = -1. \end{cases}$$

9. What are the values of $x$ in the equation,

$$2ax - x^2 = -2ab - b^2?$$

$$Ans. \begin{cases} x' = 2a + b. \\ x'' = -b. \end{cases}$$

10. What are the values of $x$ in the equation,

$$a^2 + b^2 - 2bx + x^2 = \frac{m^2x^2}{n^2}?$$

$$Ans. \begin{cases} x' = \dfrac{n}{n^2 - m^2}\Big(bn + \sqrt{a^2m^2 + b^2m^2 - a^2n^2}\Big). \\ x'' = \dfrac{n}{n^2 - m^2}\Big(bn - \sqrt{a^2m^2 + b^2m^2 - a^2n^2}\Big). \end{cases}$$

### EXAMPLES OF THE THIRD FORM.

1. What are the values of $x$ in the equation,

$$x^2 + 4x = -3?$$

First, by completing the square, we have,

$$x^2 + 4x + 4 = -3 + 4 = 1;$$

and by extracting the square root,

$$x + 2 = \pm\sqrt{1} = +1, \text{ and } -1;$$

hence, $x' = -2 + 1 = -1$; and $x'' = -2 - 1 = -3$.

That is, in this form both the roots are negative.

### VERIFICATION.

If we take the first negative value, the equation,

$$x^2 + 4x = -3,$$

gives $\qquad (-1)^2 + 4(-1) = 1 - 4 = -3$;

and by taking the second value, the equation,

$$x^2 + 4x = -3,$$

gives        $(-3)^2 + 4(-3) = 9 - 12 = -3$;

hence, both values of $x$ satisfy the given equation.

2. What are the values of $x$ in the equation,

$$-\frac{x^2}{2} - 5x - 16 = 12 + \frac{1}{2}x^2 + 6x\,?$$

By transposing and reducing, we have,

$$-x^2 - 11x = 28\,;$$

then, dividing by $-1$, the coefficient of $x^2$, we have,

$$x^2 + 11x = -28\,;$$

then, by completing the square,

$$x^2 + 11x + 30.25 = 2.25\,;$$

hence,    $x + 5.5 = \pm\sqrt{2.25} = +1.5$, and $-1.5$;

consequently,    $x' = -4$, and $x'' = -7$.

3. What are the values of $x$ in the equation,

$$-\frac{x^2}{8} - 2x - 5 = \frac{7}{8}x^2 + 5x + 5\,?$$

$$Ans. \begin{cases} x' = -2. \\ x'' = -5. \end{cases}$$

4. What are the values of $x$ in the equation,

$$2x^2 + 8x = -2\tfrac{2}{3} - \frac{2}{3}x\,?$$

$$Ans. \begin{cases} x' = -\tfrac{1}{3}. \\ x'' = -4. \end{cases}$$

5. What are the values of $x$ in the equation,

$$4x^2 + \frac{3}{5}x + 3x = -14x - 3\tfrac{1}{3} - 4x^2\,?$$

$$Ans. \begin{cases} x' = -\tfrac{1}{3}. \\ x'' = -2. \end{cases}$$

6. What are the values of $x$ in the equation,

$$- x^2 - 4 - \frac{3}{4}x = \frac{4x^2}{2} + 24x + 2 ?$$

Ans. $\begin{cases} x' = -\frac{1}{4}. \\ x'' = -8. \end{cases}$

7. What are the values of $x$ in the equation,

$$\frac{1}{9}x^2 + 7x + 20 = -\frac{8}{9}x^2 - 11x - 60 ?$$

Ans. $\begin{cases} x' = -8. \\ x'' = -10. \end{cases}$

8. What are the values of $x$ in the equation,

$$\frac{5}{6}x^2 - x + \frac{1}{2} = -9\tfrac{1}{4}x - \frac{1}{6}x^2 - \frac{1}{2} ?$$

Ans. $\begin{cases} x' = -\frac{1}{4}. \\ x'' = -8. \end{cases}$

9. What are the values of $x$ in the equation,

$$\frac{4}{5}x^2 + 5x + \frac{1}{4} = -\frac{1}{5}x^2 - 5\tfrac{1}{16}x - \frac{3}{4} ?$$

Ans. $\begin{cases} x' = -\frac{1}{16}. \\ x'' = -10. \end{cases}$

10. What are the values of $x$ in the equation,

$$x - x^2 - 3 = 6x + 1 ?$$

Ans. $\begin{cases} x' = -1. \\ x'' = -4. \end{cases}$

11. What are the values of $x$ in the equation,

$$x^2 + 4x - 90 = -93 ?$$

Ans. $\begin{cases} x' = -1. \\ x'' = -3. \end{cases}$

### EXAMPLES OF THE FOURTH FORM.

1. What are the values of $x$ in the equation,

$$x^2 - 8x = -7 ?$$

By completing the square, we have,

$$x^2 - 8x + 16 = -7 + 16 = 9;$$

then, by extracting the square root,

$$x - 4 = \pm\sqrt{9} = +3, \text{ and } -3;$$

hence,        $x' = +7,$ and $x'' = +1.$

That is, in this form, both the roots are positive.

VERIFICATION.

If we take the greater root, the equation,

$x^2 - 8x = -7,$   gives,   $7^2 - 8 \times 7 = 49 - 56 := -7;$

and for the lesser, the equation,

$x^2 - 8x = -7,$   gives,   $1^2 - 8 \times 1 = 1 - 8 = -7;$

hence, both of the roots will satisfy the equation.

2. What are the values of $x$ in the equation,

$$-1\tfrac{1}{2}x^2 + 3x - 10 = 1\tfrac{1}{2}x^2 - 18x + \frac{40}{2}?$$

By clearing of fractions, we have,

$$-3x^2 + 6x - 20 = 3x^2 - 36x + 40;$$

then, by collecting the similar terms,

$$-6x^2 + 42x = 60;$$

then, by dividing by the coefficient of $x^2$, which is $-6,$ we have,

$$x^2 - 7x = -10.$$

By completing the square, we have,

$$x^2 - 7x + 12.25 = 2.25,$$

and by extracting the square root of both members,

$$x - 3.5 = \pm\sqrt{2.25} = + 1.5, \text{ and } - 1.5;$$

hence,

$$x' = 3.5 + 1.5 = 5, \text{ and } x'' = 3.5 - 1.5 = 2.$$

### VERIFICATION.

If we take the greater root, the equation.

$$x^2 - 7x = -10, \text{ gives, } 5^2 - 7 \times 5 = 25 - 35 = -10;$$

and if we take the lesser root, the equation,

$$x^2 - 7x = -10, \text{ gives, } 2^2 - 7 \times 2 = 4 - 14 = -10.$$

3. What are the values of $x$ in the equation,

$$-3x + 2x^2 + 1 = 17\tfrac{1}{4}x - 2x^2 - 3?$$

By transposing and collecting the terms, we have,

$$4x^2 - 20\tfrac{1}{4}x = -4;$$

then dividing by the coefficient of $x^2$, we have,

$$x^2 - 5\tfrac{1}{4}x = -1.$$

By completing the square, we obtain,

$$x^2 - 5\tfrac{1}{4}x + \frac{169}{25} = -1 + \frac{169}{25} = \frac{144}{25};$$

and by extracting the root,

$$x^2 - 2\tfrac{3}{5} = \pm\sqrt{\frac{144}{25}} = + \frac{12}{5}, \text{ and } - \frac{12}{5};$$

hence,

$$x' = 2\tfrac{3}{5} + \frac{12}{5} = 5, \text{ and, } x'' = 2\tfrac{3}{5} - \frac{12}{5} = \frac{1}{5}.$$

### VERIFICATION.

If we take the greater root, the equation,

$$x^2 - 5\tfrac{1}{4}x = -1, \text{ gives, } 5^2 - 5\tfrac{1}{4} \times 5 = 25 - 26 = -1.$$

and if we take the lesser root, the equation,

$$x^2 - 5\tfrac{1}{5}x = -1, \text{ gives, } \left(\tfrac{1}{5}\right)^2 - 5\tfrac{1}{5} \times \tfrac{1}{5} = \tfrac{1}{25} - \tfrac{26}{25} = -1.$$

4. What are the values of $x$ in the equation,

$$\tfrac{1}{7}x^2 - 3x + \tfrac{1}{2} = -\tfrac{6}{7}x^2 + \tfrac{1}{4}x - \tfrac{1}{4}?$$

$$\textit{Ans. } \begin{cases} x' = 3. \\ x'' = \tfrac{1}{4}. \end{cases}$$

5. What are the values of $x$ in the equation,

$$-4x^2 - \tfrac{1}{7}x + 1\tfrac{4}{7} = -5x^2 + 8x?$$

$$\textit{Ans. } \begin{cases} x' = 8. \\ x'' = \tfrac{1}{7}. \end{cases}$$

6. What are the values of $x$ in the equation,

$$-4x^2 + \tfrac{8}{20}x - \tfrac{1}{40} = -3x^2 - \tfrac{1}{20}x + \tfrac{1}{40}?$$

$$\textit{Ans. } \begin{cases} x' = \tfrac{1}{4}. \\ x'' = \tfrac{1}{4}. \end{cases}$$

7. What are the values of $x$ in the equation,

$$x^2 - 10\tfrac{1}{10}x = -1?$$

$$\textit{Ans. } \begin{cases} x' = 10. \\ x'' = \tfrac{1}{10}. \end{cases}$$

8. What are the values of $x$ in the equation,

$$-27x + \tfrac{17x^2}{5} + 100 = \tfrac{2x^2}{5} + 12x - 26?$$

$$\textit{Ans. } \begin{cases} x' = 7. \\ x'' = 6. \end{cases}$$

9. What are the values of $x$ in the equation,

$$\tfrac{8x^2}{8} - 22x + 15 = -\tfrac{7x^2}{3} + 28x - 30?$$

$$\textit{Ans. } \begin{cases} x' = 9. \\ x'' = 1. \end{cases}$$

10. What are the values of $x$ in the equation,

$$2x^2 - 30x + 3 = -x^2 + 3\tfrac{2}{15}x - \frac{3}{10}?$$

$$Ans. \begin{cases} x'. = 11. \\ x'' = \tfrac{1}{15}. \end{cases}$$

---

## PROPERTIES OF EQUATIONS OF THE SECOND DEGREE.

### FIRST PROPERTY.

**161.** We have seen (Art. 153), that every complete equation of the second degree may be reduced to the form,

$$x^2 + 2px = q. \quad \cdots \quad (1.)$$

Completing the square, we have,

$$x^2 + 2px + p^2 = q + p^2;$$

transposing $q + p^2$ to the first member,

$$x^2 + 2px + p^2 - (q + p^2) = 0. \quad . \quad (2.)$$

Now, since $x^2 + 2px + p^2$ is the square of $x + p$, and $q + p^2$ the square of $\sqrt{q + p^2}$, we may regard the first member as the difference between two squares. Factoring, (Art. 56), we have,

$$(x + p + \sqrt{q + p^2})(x + p - \sqrt{q + p^2}) = 0. \quad . \quad (3.)$$

This equation can be satisfied only in two ways:

1st. By attributing such a value to $x$ as shall render the first factor equal to 0 ; or,

---

161. To what form may every equation of the second degree be reduced? What form will this equation take after completing the square and transposing to the first member? After factoring? In how many ways may Equation ( 3 ) be satisfied? What are they? How many roots has every equation of the second degree?

2d. By attributing such a value to $x$ as shall render the second factor equal to 0.

Placing the second factor equal to 0, we have,

$$x+p-\sqrt{q+p^2}=0; \text{ and } x'=-p+\sqrt{q+p^2}. \quad (4.)$$

Placing the first factor equal to 0, we have,

$$x+p+\sqrt{q+p^2}=0; \text{ and } x''=-p-\sqrt{q+p^2}. \quad (5.)$$

Since every supposition that will satisfy Equation ( 3 ), will also satisfy Equation ( 1 ), from which it was derived, it follows, that $x'$ and $x''$ are roots of Equation ( 1 ); also, that

*Every equation of the second degree has two roots, and only two.*

NOTE.—The two roots denoted by $x'$ and $x''$, are the same as found in Art. 158.

SECOND PROPERTY.

**162.** We have seen (Art. 161), that every equation of the second degree may be placed under the form,

$$(x+p+\sqrt{q+p^2})(x+p-\sqrt{q+p^2})=0.$$

By examining this equation, we see that the first factor may be obtained by subtracting the *second* root from the unknown quantity $x$; and the second factor by subtracting the *first* root from the unknown quantity $x$; hence,

*Every equation of the second degree may be resolved into two binomial factors of the first degree, the first terms, in both factors, being the unknown quantity, and the second terms, the roots of the equation, taken with contrary signs.*

---

162. Into how many binomial factors of the first degree may every equation of the second degree be resolved? What are the first terms of these factors? What the second?

### THIRD PROPERTY.

**163.** If we add Equations (4) and (5), Art. 161, we have,

$$x' = -p + \sqrt{q + p^2}$$
$$x'' = -p - \sqrt{q + p^2}$$

$$\overline{x' + x'' = -2p};\qquad \text{that is,}$$

*In every reduced equation of the second degree, the sum of the two roots is equal to the coefficient of the second term, taken with a contrary sign.*

### FOURTH PROPERTY.

**164.** If we multiply Equations (4) and (5), Art. 161, member by member, we have,

$$x' \times x'' = (-p + \sqrt{q + p^2})(-p - \sqrt{q + p^2})$$
$$= p^2 - (q + p^2) = -q;\quad \text{that is,}$$

*In every equation of the second degree, the product of the two roots is equal to the known term in the second member, taken with a contrary sign.*

---

### FORMATION OF EQUATIONS OF THE SECOND DEGREE.

**165.** By taking the converse of the second property, (Art. 162), we can form equations which shall have given roots; that is, if they are known, we can find the corresponding equations by the following

### RULE.

I. *Subtract each root from the unknown quantity:*

---

163. What is the algebraic sum of the roots equal to in every equation of the second degree?

164. What is the product of the roots equal to?

165. How will you find the equation when the roots are known?

II. *Multiply the results together, and place their product equal to* 0.

## EXAMPLES.

NOTE.—Let the pupil prove, *in every case*, that the roots will satisfy the third and fourth properties.

1. If the roots of an equation are 4 and $-5$, what is the equation?                    *Ans.* $x^2 + x = 20.$

2. What is the equation when the roots are 1 and $-3$?
                              *Ans.* $x^2 + 2x = 3.$

3. What is the equation when the roots are 9 and $-10$?
                              *Ans.* $x^2 + x = 90.$

4. What is the equation whose roots are 6 and $-10$?
                              *Ans.* $x^2 + 4x = 60.$

5. What is the equation whose roots are 4 and $-3$?
                              *Ans.* $x^2 - x = 12.$

6. What is the equation whose roots are 10 and $-\frac{1}{10}$?
                              *Ans.* $x^2 - 9\frac{9}{10}x = 1.$

7. What is the equation whose roots are 8 and $-2$?
                              *Ans.* $x^2 - 6x = 16.$ .

8. What is the equation whose roots are 16 and $-5$?
                              *Ans.* $x^2 - 11x = 80.$

9. What is the equation whose roots are $-4$ and $-5$?
                              *Ans.* $x^2 + 9x = -20.$

10. What is the equation whose roots are $-6$ and $-7$?
                              *Ans.* $x^2 + 13x = -42.$

11. What is the equation whose roots are $-\frac{3}{4}$ and $-2$?

                    *Ans.* $x^2 + 2\frac{3}{4}x = -\frac{3}{2}.$

12. What is the equation whose roots are $-2$ and $-3$?
                              *Ans.* $x^2 + 5x = -6.$

13. What is the equation whose roots are 4 and 3?

$$Ans.\ x^2 - 7x = -12.$$

14. What is the equation whose roots are 12 and 2?

$$Ans.\ x^2 - 14x = -24.$$

15. What is the equation whose roots are 18 and 2?

$$Ans.\ x^2 - 20x = -36.$$

16. What is the equation whose roots are 14 and 3?

$$Ans.\ x^2 - 17x = -42.$$

17. What is the equation whose roots are $\dfrac{4}{9}$ and $-\dfrac{9}{4}$?

$$Ans.\ x^2 + \frac{65}{36}x = 1.$$

18. What is the equation whose roots are 5 and $-\dfrac{2}{3}$?

$$Ans.\ x^2 - \frac{13}{3}x = \frac{10}{3}.$$

19. What is the equation whose roots are $a$ and $b$?

$$Ans.\ x^2 - (a + b)x = -ab.$$

20. What is the equation whose roots are $c$ and $-d$?

$$Ans.\ x^2 - (c - d)x = cd.$$

---

## TRINOMIAL EQUATIONS OF THE SECOND DEGREE.

**165.** A trinomial equation of the second degree contains three kinds of terms:

1st. A term involving the unknown quantity to the second degree.

2d. A term involving the unknown quantity to the first degree; and

3d. A known term. Thus,

$$x^2 - 4x - 12 = 0,$$

is a trinomial equation of the second degree.

FACTORING.

**165.**\*\* What are the factors of the trinomial equation,

$$x^2 - 4x - 12 = 0 ?$$

A trinomial equation of the second degree may always be reduced to one of the four forms (Art. 160), by simply transposing the known term to the second member, and then solving the equation. Thus, from the above equation, we have,

$$x^2 - 4x = 12.$$

Resolving the equation, we find the two roots to be $+ 6$ and $- 2$; therefore, the factors are, $x - 6$, and $x + 2$ (Art. 162).

Since the sum of the two roots is equal to the coefficient of the second term, taken with a contrary sign (Art. 163); and the product of the two roots is equal to the known term in the second member, taken with a contrary sign, or to the third term of the trinomial, taken with the *same sign:* hence it follows, that any trinomial may be factored by inspection, when two numbers can be discovered *whose algebraic sum is equal to the coefficient of the second term, and whose product is equal to the third term.*

EXAMPLES

1. What are the factors of the trinomial, $x^2 - 9x - 36$ ?

It is seen, by inspection, that $- 12$ and $+ 3$ will fulfil the conditions of roots. For, $12 - 3 = 9$; that is, the coefficient of the second term with a contrary sign; and $12 \times - 3 = - 36$, the third term of the trinomial; hence, the factors are, $x - 12$, and $x + 3$.

2. What are the factors of $x^2 - 7x - 30 = 0$ ?

*Ans.* $x - 10$, and $x + 3$

**3.** What are the factors of $x^2 + 15x + 36 = 0$?

*Ans.* $x + 12$, and $x + 3$.

**4.** What are the factors of $x^2 - 12x - 28 = 0$?

*Ans.* $x - 14$, and $x + 2$.

**5.** What are the factors of $x^2 - 7x - 8 = 0$?

*Ans.* $x - 8$, and $x + 1$.

### TRINOMIAL EQUATIONS OF THE FORM

$$x^{2n} + 2px^n = q.$$

In the above equation, the exponent of $x$, in the first term, is double the exponent of $x$ in the second term.

$$x^6 - 4x^3 = 32, \quad \text{and} \quad x^4 + 4x^2 = 117,$$

are both equations of this form, and may be solved by the rules already given for the solution of equations of the second degree.

In the equation,

$$x^{2n} + 2px^n = q,$$

we see that the first member will become a perfect square, by adding to it the square of half the coefficient of $x^n$; thus,

$$x^{2n} + 2px^n + p^2 = q + p^2,$$

in which the first member is a perfect square. Then, extracting the square root of both members, we have,

$$x^n + p = \pm \sqrt{q + p^2};$$

hence, $\qquad x^n = -p \pm \sqrt{q + p^2};$

then, by taking the $n$th root of both members,

$$x' = \sqrt[n]{-p + \sqrt{q + p^2}},$$

and $\qquad x'' = \sqrt[n]{-p - \sqrt{-p + p^2}}.$

## EXAMPLES.

1. What are the values of $x$ in the equation,

$$x^6 + 6x^3 = 112?$$

Completing the square,

$$x^6 + 6x^3 + 9 = 112 + 9 = 121;$$

then, extracting the square root of both members,

$$x^3 + 3 = \pm \sqrt{121} = \pm 11; \text{ hence,}$$

$$x' = \sqrt[3]{-3 + 11}, \quad \text{and} \quad x'' = \sqrt[3]{-3 - 11}; \text{ hence,}$$

$$x' = \sqrt[3]{8} = 2, \quad \text{and} \quad x'' = \sqrt[3]{-14} = -\sqrt[3]{14}.$$

2. What are the values of $x$ in the equation,

$$x^4 - 8x^2 = 9?$$

Completing the square, we have,

$$x^4 - 8x^2 + 16 = 9 + 16 = 25.$$

Extracting the square root of both members,

$$x^2 - 4 = \pm \sqrt{25} = \pm 5; \text{ hence,}$$

$$x' = \pm \sqrt{4 + 5}, \quad \text{and} \quad x'' = \pm \sqrt{4 - 5}; \text{ hence,}$$

$$x' = +3 \text{ and } -3; \quad \text{and} \quad x'' = + \sqrt{-1} \text{ and } - \sqrt{-1}.$$

3. What are the values of $x$ in the equation,

$$x^6 + 20x^3 = 69?$$

Completing the square,

$$x^6 + 20x^3 + 100 = 69 + 100 = 169.$$

Extracting the square root of both members,

$$x^3 + 10 = \pm \sqrt{169} = \pm 13; \text{ hence,}$$

$$x' = \sqrt[3]{-10 + 13}, \quad \text{and} \quad x'' = \sqrt[3]{-10 - 13}.$$

$$x' = \sqrt[3]{3}, \quad \text{and} \quad x'' = \sqrt[3]{-23}.$$

4. What are the values of $x$ in the equation,

$$x^4 - 2x^2 = 3 ?$$

*Ans.* $x' = \pm \sqrt{3}$, and $x'' = \pm \sqrt{-1}$.

5. What are the values of $x$ in the equation,

$$x^6 + 8x^3 = 9 ?$$

*Ans.* $x' = 1$, and $x'' = \sqrt[3]{-9}$.

6. Given $x + \sqrt{9x + 4} = 12$, to find $x$.

Transposing $x$ to the second member, and then squaring,

$$9x + 4 = x^2 - 24x + 144 ;$$

$$\therefore \quad x^2 - 33x = -140 ;$$

and,         $x' = 28$,   and   $x'' = 5$.

7. $4x + 4\sqrt{x + 2} = 7$.     *Ans.* $x' = 4\frac{1}{4}$, $x'' = \frac{1}{4}$.

8. $x + \sqrt{5x + 10} = 8$.     *Ans.* $x' = 18$, $x'' = 3$.

———

## NUMERICAL VALUES OF THE ROOTS.

**166.** We have seen (Art. 160), that by attributing all possible signs to $2p$ and $q$, we have the four following forms:

$$x^2 + 2px = q. \quad \ldots \ldots (1.)$$
$$x^2 - 2px = q. \quad \ldots \ldots (2.)$$
$$x^2 + 2px = -q. \quad \ldots \ldots (3.)$$
$$x^2 - 2px = -q. \quad \ldots \ldots (4.)$$

166. To how many forms may every equation of the second degree be reduced? What are they?

## First Form.

**167.** Since $q$ is positive, we know, from Property Fourth, that the product of the roots must be negative; hence, *the roots have contrary signs.* Since the coefficient $2p$ is positive, we know, from Property Third, that the algebraic sum of the roots is negative; hence, *the negative root is numerically the greater.*

## Second Form.

**168.** Since $q$ is positive, the product of the roots must be negative; hence, *the roots have contrary signs.* Since $2p$ is negative, the algebraic sum of the roots must be positive; hence, *the positive root is numerically the greater.*

## Third Form.

**169.** Since $q$ is negative, the product of the roots is positive (Property Fourth); hence, *the roots have the same sign.* Since $2p$ is positive, the sum of the roots must be negative; hence, *both are negative.*

## Fourth Form.

**170.** Since $q$ is negative, the product of the roots is positive; hence, *the roots have the same sign.* Since $2p$ is negative, the sum of the roots is positive; hence, *the roots are both positive.*

---

167. What sign has the product of the roots in the first form? How are their signs? Which root is numerically the greater? Why?

168. What sign has the product of the roots in the second form? How are the signs of the roots? Which root is numerically the greater?

169. What sign has the product of the roots in the third form? How are their signs?

170. What sign has the product of the roots in the fourth form? How are the signs of the roots?

## First and Second Forms.

**171.** If we make $q = 0$, the first form becomes,

$$x^2 + 2px = 0, \quad \text{or} \quad x(x + 2p) = 0;$$

which shows that one root is equal to 0, and the other to $-2p$.

Under the same supposition, the second form becomes,

$$x^2 - 2px = 0, \quad \text{or} \quad x(x - 2p) = 0;$$

which shows that one root is equal to 0, and the other to $2p$. Both of these results are as they should be; since, when $q$, the product of the roots, becomes 0, one of the factors must be 0; and hence, one root must be 0.

## Third and Fourth Forms.

**172.** If, in the Third and Fourth Forms, $q > p^2$, the quantity under the radical sign will become *negative ;* hence, *its square root cannot be extracted* (Art. 137). Under this supposition, the values of $x$ are *imaginary.* How are these results to be interpreted?

*If a given number be divided into two parts, their product will be the greatest possible, when the parts are equal.*

Denote the number by $2p$, and the difference of the parts by $d$; then,

$$p + \frac{d}{2} = \text{the greater part,} \quad \text{(Page 120.)}$$

and, $$p - \frac{d}{2} = \text{the less part,}$$

and, $$p^2 - \frac{d^2}{4} = P, \text{ their product.}$$

171. If we make $q = 0$, to what does the first form reduce? What, then, are its roots? Under the same supposition, to what does the second form reduce? What are, then, its roots?

172. If $q > p^2$, in the third and fourth forms, what takes place?

If a number be divided into two parts, when will the product be the greatest possible?

It is plain, that the product $P$ will *increase*, as $d$ *diminishes*, and that it will be the *greatest possible* when $d = 0$; for then there will be no negative quantity to be subtracted from $p^2$, in the first member of the equation. But when $d = 0$, the parts are equal; hence, *the product of the two parts is the greatest when they are equal.*

In the equations,

$$x^2 + 2px = -q, \qquad x^2 - 2px = -q,$$

$2p$ is the sum of the roots, and $-q$ their product; and hence, by the principle just established, the product $q$, can never be greater than $p^2$. This condition fixes a limit to the value of $q$. If, then, we make $q > p^2$, we pass this limit, and express, by the equation, a condition which cannot be fulfilled; and this incompatibility of the conditions is made apparent by the values of $x$ becoming imaginary. Hence, we conclude that,

*When the values of the unknown quantity are imaginary, the conditions of the proposition are incompatible with each other.*

### EXAMPLES.

1. Find two numbers, whose sum shall be 12 and product 46.

Let $x$ and $y$ be the numbers.

By the 1st condition, $x + y = 12$;

and by the 2d, $\qquad\qquad xy = 46.$

The first equation gives,

$$x = 12 - y.$$

Substituting this value for $x$ in the second, we have,

$$12y - y^2 = 46;$$

and changing the signs of the terms, we have,

$$y^2 - 12y = -46$$

Then, by completing the square,

$$y^2 - 12y + 36 = -46 + 36 = -10;$$

which gives,     $y' = 6 + \sqrt{-10}$,

and,     $y'' = 6 - \sqrt{-10}$;

both of which values are imaginary, as indeed they should be, since the conditions are incompatible.

2. The sum of two numbers is 8, and their product 20: what are the numbers?

Denote the numbers by $x$ and $y$.

By the first condition,

$$x + y = 8;$$

and by the second,     $xy = 20$.

The first equation gives,

$$x = 8 - y.$$

Substituting this value of $x$ in the second, we have,

$$8y - y^2 = 20;$$

changing the signs, and completing the square, we have,

$$y^2 - 8y + 16 = -4;$$

and by extracting the root,

$$y' = 4 + \sqrt{-4}, \text{ and } y'' = 4 - \sqrt{-4}.$$

These values of $y$ may be put under the forms (Art. 142),

$$y = 4 + 2\sqrt{-1}, \text{ and } y = 4 - 2\sqrt{-1}.$$

3 What are the values of $x$ in the equation,

$$x^2 + 2x = -10?$$

$$Ans. \begin{cases} x' = -1 + 3\sqrt{-1}. \\ x'' = -1 - 3\sqrt{-1}. \end{cases}$$

## PROBLEMS.

1. Find a number such, that twice its square, added to three times the number, shall give 65.

Let $x$ denote the unknown number. Then, the equation of the problem is,

$$2x^2 + 3x = 65 \; ;$$

whence,

$$x = -\frac{3}{4} \pm \sqrt{\frac{65}{2} + \frac{9}{16}} = -\frac{3}{4} \pm \frac{23}{4}.$$

Therefore,

$$x' = -\frac{3}{4} + \frac{23}{4} = 5, \quad \text{and} \quad x'' = -\frac{3}{4} - \frac{23}{4} = -\frac{13}{2}.$$

Both these values satisfy the equation of the problem. For,

$$2 \times (5)^2 + 3 \times 5 = 2 \times 25 + 15 = 65 \; ;$$

and, $\quad 2\left(-\frac{13}{2}\right)^2 + 3 \times -\frac{13}{2} = \frac{169}{2} - \frac{39}{2} = \frac{130}{2} = 65.$

Notes.—1. If we restrict the enunciation of the problem to its arithmetical sense, in which "added" means *numerical increase*, the first value of $x$ only will satisfy the conditions of the problem.

2. If we give to "added," its algebraical signification (when it may mean subtraction as well as addition), the problem may be thus stated :

To find a number such, that twice its square diminished by three times the number, shall give 65.

The second value of $x$ will satisfy this enunciation; for,

$$2\left(\frac{13}{2}\right)^2 - 3 \times \frac{13}{2} = \frac{169}{2} - \frac{39}{2} = 65.$$

3. The root which results from giving the plus sign to the radical, is, generally, an answer to the question in its arithmetical sense. The second root generally satisfies the problem under a modified statement.

Thus, in the example, it was required to find a number, of which twice the square, *added* to three times the number, shall give 65. Now, in the arithmetical sense, *added* means increased; but in the algebraic sense, it implies diminution when the quantity added is negative. In this sense, the second root satisfies the enunciation.

2. A certain person purchased a number of yards of cloth for 240 cents. If he had purchased 3 yards *less* of the same cloth for the same sum, it would have cost him 4 cents more per yard: how many yards did he buy?

Let $x$ denote the number of yards purchased.

Then, $\dfrac{240}{x}$ will denote the price per yard.

If, for 240 cents, he had purchased three yards less, that is, $x - 3$ yards, the price per yard, under this hypothesis, would have been denoted by $\dfrac{240}{x - 3}$. But, by the conditions, this last cost must exceed the first by 4 cents. Therefore, we have the equation,

$$\frac{240}{x - 3} - \frac{240}{x} = 4;$$

whence, by reducing, $x^2 - 3x = 180$,

and, $x = \dfrac{3}{2} \pm \sqrt{\dfrac{9}{4} + 180} = \dfrac{3 \pm 27}{2};$

therefore, $x' = 15$, and $x'' = -12$.

Notes.—1. The value, $x' = 15$, satisfies the enunciation in its arithmetical sense. For, if 15 yards cost 240 cents,

11

$240 \div 15 = 16$ cents, the price of 1 yard; and $240 \div 12 = 20$ cents, the price of 1 yard under the second supposition.

2. The second value of $x$ is an answer to the following Problem:

A certain person purchased a number of yards of cloth for 240 cents. If he had paid the same for three yards *more*, it would have cost him 4 cents *less* per yard: how many yards did he buy?

This would give the equation of condition,

$$\frac{240}{x} - \frac{240}{x + 3} = 4; \text{ or,}$$

$$x^2 - 3x = 180;$$

the same equation as found before; hence,

*A single equation will often state two or more arithmetical problems.*

This arises from the fact that the language of Algebra is more comprehensive than that of Arithmetic.

3. A man having bought a horse, sold it for $24. At the sale he lost as much per cent. on the price of the horse, as the horse cost him dollars: what did he pay for the horse?

Let $x$ denote the number of dollars that he paid for the horse. Then, $x - 24$ will denote the loss he sustained. But as he lost $x$ per cent. by the sale, he must have lost $\frac{x}{100}$ upon each dollar, and upon $x$ dollars he lost a sum denoted by $\frac{x^2}{100}$; we have, then, the equation,

$$\frac{x^2}{100} = x - 24; \text{ whence, } x^2 - 100x = -2400,$$

and,    $x = 50 \pm \sqrt{2500 - 2400} = 50 \pm 10.$

Therefore,    $x' = 60$,  and  $x'' = 40.$

Both of these roots will satisfy the problem.

For, if the man gave $60 for the horse, and sold him for $24, he lost $36. From the enunciation, he should have lost 60 per cent. of $60 ; that is,

$$\frac{60}{100} \text{ of } 60 = \frac{60 \times 60}{100} = 36 ;$$

therefore, $60 satisfies the enunciation.

Had he paid $40 for the horse, he would have lost by the sale, $16. From the enunciation, he should have lost 40 per cent. of $40 ; that is,

$$\frac{40}{100} \text{ of } 40 = \frac{40 \times 40}{100} = 16 ;$$

therefore, $40 satisfies the enunciation.

4. The sum of two numbers is 11, and the sum of their squares is 61 : what are the numbers?    *Ans.* 5 and 6.

5. The difference of two numbers is 3, and the sum of their squares is 89 : what are the numbers?    *Ans.* 5 and 8.

6. A grazier bought as many sheep as cost him £60, and after reserving fifteen out of the number, he sold the remainder for £54, and gained 2s. a head on those he sold : how many did he buy?    *Ans.* 75.

7. A merchant bought cloth, for which he paid £33 15s., which he sold again at £2 8s. per piece, and gained by the bargain as much as one piece cost him: how many pieces did he buy?    *Ans.* 15.

8. The difference of two numbers is 9, and their sum, multiplied by the greater, is equal to 266 : what are the numbers?    *Ans.* 14 and 5.

9. To find a number, such that if you subtract it from 10, and multiply the remainder by the number itself, the product will be 21.                    *Ans.* 7 or 3.

10. A person traveled 105 miles. If he had traveled 2 miles an hour slower, he would have been 6 hours longer in completing the same distance: how many miles did he travel per hour?                    *Ans.* 7 miles.

11. A person purchased a number of sheep, for which he paid $224. Had he paid for each twice as much, plus 2 dollars, the number bought would have been denoted by twice what was paid for each : how many sheep were purchased?
                    *Ans.* 32.

12. The difference of two numbers is 7, and their sum multiplied by the greater, is equal to 130: what are the numbers?                    *Ans.* 10 and 3.

13. Divide 100 into two such parts, that the sum of their squares shall be 5392.                    *Ans.* 64 and 36.

14. Two square courts are paved with stones a foot square; the larger court is 12 feet larger than the smaller one, and the number of stones in both pavements is 2120 : how long is the smaller pavement?                    *Ans.* 26 feet.

15. Two hundred and forty dollars are equally distributed among a certain number of persons. The same sum is again distributed amongst a number greater by 4. In the latter case each receives 10 dollars less than in the former: how many persons were there in each case.        *Ans.* 8 and 12.

16. Two partners, *A* and *B*, gained 360 dollars. *A's* money was in trade 12 months, and he received, for principal and profit, 520 dollars. *B's* money was 600 dollars, and was in trade 16 months: how much capital had *A* ?
                    *Ans.* 400 dollars.

**173.** Two simultaneous equations, each of the second degree, and containing two unknown quantities, will, when combined, generally give rise to an equation of the fourth degree. Hence, only particular cases of such equations can be solved by the methods already given.

FIRST.

*Two simultaneous equations, involving two unknown quantities, can always be solved when one is of the first and the other of the second degree.*

EXAMPLES.

1. Given $\begin{cases} x + y = 14 \\ x^2 + y^2 = 100 \end{cases}$ to find $x$ and $y$.

By transposing $y$ in the first equation, we have,

$$x = 14 - y;$$

and by squaring both members,

$$x^2 = 196 - 28y + y^2.$$

Substituting this value for $x^2$ in the second equation, we have,

$$196 - 28y + y^2 + y^2 = 100;$$

from which we have,

$$y^2 - 14y = -48.$$

By completing the square,

$$y^2 - 14y + 49 = 1;$$

---

173. When may two simultaneous equations of the second degree be solved?

and by extracting the square root,

$$y - 7 = \pm\sqrt{1} = +1, \quad \text{and} \quad -1;$$

hence, $y' = 7 + 1 = 8,$ and $y'' = 7 - .1 = 6.$

If we take the greater value, we find $x = 6$; and if we take the lesser, we find $x = 8.$

$$Ans. \begin{cases} x' = 8, & x'' = 6. \\ y' = 6, & y'' = 8 \end{cases}$$

### VERIFICATION.

For the greater value, $y = 8,$ the equation,

$$x + y = 14, \quad \text{gives} \quad 6 + 8 = 14;$$

and, $x^2 + y^2 = 100,$ gives $36 + 64 = 100.$

For the value $y = 6,$ the equation,

$$x + y = 14, \quad \text{gives} \quad 8 + 6 = 14;$$

and, $x^2 + y^2 = 100,$ gives $64 + 36 = 100.$

Hence, both sets of values satisfy the given equation.

2. Given $\begin{cases} x - y = 3 \\ x^2 - y^2 = 45 \end{cases}$ to find $x$ and $y.$

Transposing $y$ in the first equation, we have,

$$x = 3 + y;$$

then, squaring both members,

$$x^2 = 9 + 6y + y^2.$$

Substituting this value for $x^2,$ in the second equation, we have,

$$9 + 6y + y^2 - y^2 = 45;$$

whence, we have,

$$6y = 36, \quad \text{and} \quad y = 6.$$

Substituting this value of $y$, in the first equation, we have,

$$x - 6 = 3,$$

and, consequently,   $x' = 3 + 6 = 9.$

### VERIFICATION.

$$x - y = 3, \text{ gives } 9 - 6 = 3;$$

and,     $x^2 - y^2 = 45$, gives $81 - 36 = 45.$

Solve the following simultaneous equations:

3. $\begin{cases} x + y = 12 \\ x^2 - y^2 = 24 \end{cases}$        *Ans.* $\begin{cases} x' = 7. \\ y' = 5. \end{cases}$

4. $\begin{cases} x - y = 3 \\ x^2 + y^2 = 117 \end{cases}$    *Ans.* $\begin{cases} x' = 9, \quad x'' = -6. \\ y' = 6, \quad y'' = -9. \end{cases}$

5. $\begin{cases} x + y = 9 \\ x^2 - 2xy + y^2 = 1 \end{cases}$    *Ans.* $\begin{cases} x' = 5, \quad '' = 5. \\ y' = 4, \quad y'' = 4. \end{cases}$

6. $\begin{cases} x - y = 5 \\ x^2 + 2xy + y^2 = 225 \end{cases}$

$\qquad\qquad$ *Ans.* $\begin{cases} x' = 10, \quad x'' = -5. \\ y' = 5, \quad y'' = -10. \end{cases}$

### SECOND.

**174.** *Two simultaneous equations of the second degree, which are homogeneous with respect to the unknown quantity, can always be solved.*

### EXAMPLES.

1. Given $\begin{cases} x^2 + 3xy = 22 & \quad . \quad . \quad . \quad . \quad . \text{ (1.)} \\ x^2 + 3xy + 2y^2 = 40 & . \quad . \quad . \quad . \quad . \text{ (2.)} \end{cases}$

to find $x$ and $y$.

---

174. When may two simultaneous equations of the second degree be solved?

Assume $x = ty$, $t$ being any auxiliary unknown quantity. Substituting this value of $x$ in Equations (1) and (2), we have,

$$t^2y^2 + 3ty^2 = 22, \quad \therefore \ y^2 = \frac{22}{t^2 + 3t}; \qquad (3.)$$

$$t^2y^2 + 3ty^2 + 2y^2 = 40, \quad \therefore \ y^2 = \frac{40}{t^2 + 3t + 2}; \quad (4.)$$

hence, $\qquad \dfrac{22}{t^2 + 3t} = \dfrac{40}{t^2 + 3t + 2};$

hence, $\qquad 22t^2 + 66t + 44 = 40t^2 + 120t;$

reducing, $\qquad t^2 + 3t = \dfrac{22}{9};$

whence, $\qquad t' = \dfrac{2}{3}, \ \text{ and } \ t'' = -\dfrac{11}{3}.$

Substituting either of these values in Equations (3) or (4), we find,
$$y' = +3, \ \text{ and } \ y'' = -3$$

Substituting the plus value of $y$, in Equation (1), we have,
$$x^2 + 9x = 22;$$
from which we find,
$$x' = +2, \ \text{ and } \ x'' = -11.$$

If we take the negative value, $y'' = -3$, we have, from Equation (1),
$$x^2 - 9x = 22;$$
from which we find,
$$x' = +11, \ \text{ and } \ x'' = -2.$$

VERIFICATION.

For the values $y' = +3$, and $x' = +2$, the given equation,
$$x^2 + 3xy = 22,$$

gives, $2^2 + 3 \times 2 \times 3 = 4 + 18 = 22$;

and for the second value, $x'' = -11$, the same equation,

$$x^2 + 3xy = 22,$$

gives, $(-11)^2 + 3 \times -11 \times 3 = 121 - 99 = 22.$

If, now, we take the second value of $y$, that is, $y'' = -3$, and the corresponding values of $x$, viz., $x' = +11$, and $x'' = -2$; for $x' = +11$, the given equation,

$$x^2 + 3xy = 22,$$

gives, $11^2 + 3 \times 11 \times -3 = 121 - 99 = 22$;

and for $x'' = -2$, the same equation,

$$x^2 + 3xy = 22,$$

gives, $(-2)^2 + 3 \times -2 \times -3 = 4 + 18 = 22.$

The verifications could be made in the same way by employing Equation (2).

NOTE.—In equations similar to the above, we generally find but a single pair of values, corresponding to the values in this equation, of $y' = +3$, and $x' = +2$.

The complete solution would give four pairs of values.

2. $\left\{ \begin{array}{l} x^2 - y^2 = -9 \\ y^2 - xy = 5 \end{array} \right\}$     $Ans.$ $\left\{ \begin{array}{l} x = 4. \\ y = 5. \end{array} \right.$

3. $\left\{ \begin{array}{l} xy - y^2 = -7 \\ y^2 + x^2 = 85 \end{array} \right\}$     $Ans.$ $\left\{ \begin{array}{l} x = 6. \\ y = 7. \end{array} \right.$

4. $\left\{ \begin{array}{l} 2x^2 + 3xy = 470 \\ y^2 - xy = -9 \end{array} \right\}$     $Ans.$ $\left\{ \begin{array}{l} x = 10. \\ y = 9. \end{array} \right.$

5. $\left\{ \begin{array}{l} 5xy - 3y^2 = 32 \\ x^2 + y^2 + 3xy = 71 \end{array} \right\}$     $Ans.$ $\left\{ \begin{array}{l} x = 7. \\ y = 1. \end{array} \right.$

11*

### THIRD.—PARTICULAR CASES.

**175.** Many other equations of the second degree may be so transformed, as to be brought under the rules of solution already given. The seven following formulas will aid in such transformation.

### ( 1.)

When the sum and difference are known:

$$x + y = s$$
$$x - y = d.$$

Then, page 132, Example 3,

$$x = \frac{s + d}{2} = \frac{1}{2}s + \frac{1}{2}d, \quad \text{and} \quad y = \frac{s - d}{2} = \frac{1}{2}s - \frac{1}{2}d.$$

### ( 2.)

When the sum and product are known:

$$x + y = s \ldots \ldots \ldots \text{(1.)}$$
$$xy = p \ldots \ldots \ldots \text{(2.)}$$
$$x^2 + 2xy + y^2 = s^2, \quad \text{by squaring ( 1 );}$$
$$4xy = 4p, \quad \text{by mult. ( 2 ) by 4.}$$

$$\overline{x^2 - 2xy + y^2 = s^2 - 4p, \quad \text{by subtraction.}}$$

$$x - y = \pm \sqrt{s^2 - 4p}, \quad \text{by ext. root.}$$

But, $$\qquad x + y = s;$$

hence $$\qquad x = \frac{s}{2} \pm \frac{1}{2} \sqrt{s^2 - 4p}.$$

and, $$\qquad y = \frac{s}{2} \mp \frac{1}{2} \sqrt{s^2 - 4p}.$$

---

175. What is the first formula of this article? What the second? Third? Fourth? Fifth? Sixth? Seventh?

## (3.)

When the difference and product are known:

$$x - y = d \quad \ldots \ldots \ldots \quad (1.)$$
$$xy = p \quad \ldots \ldots \ldots \quad (2.)$$
$$x^2 - 2xy + y^2 = d^2, \quad \text{by squaring (1)} \cdot$$
$$\underline{4xy \qquad = 4p, \quad \text{mult. (2) by 4.}}$$
$$x^2 + 2xy + y^2 = d^2 + 4p, \quad \text{by adding.}$$
$$x + y = \pm \sqrt{d^2 + 4p}$$
$$\underline{x - y = d}$$
$$x = \tfrac{1}{2}d \pm \tfrac{1}{2}\sqrt{d^2 + 4p}.$$
$$y = -\tfrac{1}{2}d \pm \tfrac{1}{2}\sqrt{d^2 + 4p}.$$

## 4.)

When the sum of the squares and product are known.

$$x^2 + y^2 = s \ldots (1.) \quad xy = p \ldots (2.) \quad \therefore 2xy = 2p \ldots (3.)$$

Adding (1) and (3), $x^2 + 2xy + y^2 = s + 2p$;

hence, $\qquad\qquad x + y = \pm \sqrt{s + 2p} \quad (4.)$

Subtracting (3) from (1), $x^2 - 2xy + y^2 = s - 2p$;

hence, $\qquad\qquad x - y = \pm \sqrt{s - 2p} \quad (5.)$

Combining (4) and (5), $x = \tfrac{1}{2}\sqrt{s + 2p} + \tfrac{1}{2}\sqrt{s - 2p}$,

and, $\qquad\qquad y = \tfrac{1}{2}\sqrt{s + 2p} - \tfrac{1}{2}\sqrt{s - 2p}.$

## (5.)

When the sum and sum of the squares are known:

$$x + y = s \quad \ldots \ldots \quad (1.)$$
$$x^2 + y^2 = s' \quad \ldots \ldots \quad (2.)$$
$$\underline{x^2 + 2xy + y^2 = s^2 \quad \text{by squaring (1)}}$$
$$2xy = s^2 - s'$$
$$xy = \frac{s^2 - s'}{2} = p. \quad (3.)$$

By putting $xy = p$, and combining Equations (1) and (3), by Formula (**2**), we find the values of $x$ and $y$.

## (6.)

When the sum and sum of the cubes are known:

$$x + y = 8 \quad \cdots \cdots \quad (1.)$$
$$x^3 + y^3 = 152 \quad \cdots \cdots \quad (2.)$$
$$\underline{x^3 + 3x^2y + 3xy^2 + y^3 = 512 \quad \text{by cubing (1)}.}$$
$$3x^2y + 3xy^2 = 360 \quad \text{by subtraction}.$$
$$3xy(x + y) = 360 \quad \text{by factoring}.$$
$$3xy(8) = 360 \quad \text{from Equa. (1)}.$$
$$24xy = 360$$

hence, $\qquad\qquad xy = 15 \quad \cdots \cdots \quad (3.)$

Combining (1) and (3), we find $x = 5$ and $y = 3$

## (7.)

When we have an equation of the form,

$$(x + y)^2 + (x + y) = q.$$

Let us assume $x + y = z$.

Then the given equation becomes,

$$z^2 + z = q; \quad \text{and} \quad z = -\frac{1}{2} \pm \sqrt{q + \frac{1}{4}}.$$

$$x + y = -\frac{1}{2} \pm \sqrt{q + \frac{1}{4}}.$$

## EXAMPLES.

1. Given $\begin{cases} xz = y^2 & (1) \\ x + y + z = 7 & (2) \\ x^2 + y^2 + z^2 = 21 & (3) \end{cases}$ to find $x$, $y$, and $z$,

Transposing $y$ in Equation ( 2 ), we have,

$$x + z = 7 - y; \qquad \ldots \quad (4.)$$

then, squaring the members, we have,

$$x^2 + 2xz + z^2 = 49 - 14y + y^2.$$

If now we substitute for $2xz$, its value taken from Equation ( 1 ), we have,

$$x^2 + 2y^2 + z^2 = 49 - 14y + y^2;$$

and cancelling $y^2$, in each member, there results,

$$x^2 + y^2 + z^2 = 49 - 14\,y. \quad .$$

But, from Equation ( 3 ), we see that each member of the last equation is equal to 21 ; hence,

$$49 - 14y = 21,$$

and, $\qquad\qquad 14y = 49 - 21 = 28,$

hence, $\qquad\qquad y = \dfrac{28}{14} = 2.$

Substituting this value of $y$ in Equation ( 1 ), gives,

$$xz = 4;$$

and substituting it in Equation ( 4 ), gives,

$$x + z = 5, \quad \text{or} \quad x = 5 - z.$$

Substituting this value of $x$, in the previons equation, we obtain

$$. \; 5z - z^2 = 4, \quad \text{or} \quad z^2 - 5z = -4;$$

and by completing the square, we have,

$$z^2 - 5z + 6.25 = 2.5,$$

and, $\quad z - 2.5 = \pm\sqrt{2.5} = +1.5, \quad \text{or} \quad -1.5;$

hence, $z = 2.5 + 1.5 = 4$, and $z = +2.5 - 1.5 = 1$

2. Given $x + \sqrt{xy} + y = 19$ and $x^2 + xy + y^2 = 133$ } to find $x$ and $y$.

Dividing the second equation by the first, we have,

$$x - \sqrt{xy} + y = 7$$

but, $\qquad\qquad x + \sqrt{xy} + y = 19$

hence, by addition, $\qquad 2x + 2y = 26$

or, $\qquad\qquad\qquad x + y = 13$

and substituting in 1st Equa., $\sqrt{xy} + 13 = 19$

or, by transposing, $\qquad\qquad \sqrt{xy} = 6$

and by squaring, $\qquad\qquad xy = 36.$

Equation 2d, is $\qquad\qquad x^2 + xy + y^2 = 133$

and from the last, we have, $\qquad 3xy = 108$

Subtracting, $\qquad\qquad x^2 - 2xy + y^2 = 25$

hence, $\qquad\qquad\qquad x - y = \pm 5$

but, $\qquad\qquad\qquad\qquad x + y = 13$

hence, $\quad x = 9$ and $4$; and $y = 4$ and $9$.

———

## PROBLEMS.

1. Find two numbers, such that their sum shall be 15 and the sum of their squares 113.

Let $x$ and $y$ denote the numbers; then,

$$x + y = 15, \quad (1.) \quad \text{and} \quad x^2 + y^2 = 113. \quad (2.)$$

From Equation ( 1 ), we have,

$$x^2 = 225 - 30y + y^2$$

Substituting this value in Equation ( 2 ),

$$225 - 30y + y^2 + y^2 = 113;$$

hence, $2y^2 - 30y = -112$ ;

$$y^2 - 15y = -56, \cdot$$

hence, $y' = 8$, and $y'' = 7$.

The first value of $y$ being substituted in Equation (1), gives $x' = 7$; and the second, $x'' = 8$. Hence, the numbers are 7 and 8.

2. To find two numbers, such that their product added to their sum shall be 17, and their sum taken from the sum of their squares shall leave 22.

Let $x$ and $y$ denote the numbers; then, from the conditions,

$$(x + y) + xy = 17. \quad \cdots \quad (1.)$$

$$x^2 + y^2 - (x + y) = 22. \quad \cdots \quad (2.)$$

Multiplying Equation (1) by 2, we have,

$$2xy + 2(x + y) = 34. \quad \cdots \quad (3.)$$

Adding (2) and (3), we have,

$$x^2 + 2xy + y^2 + (x + y) = 56 ;$$

hence, $(x + y)^2 + (x + y) = 56. \quad \cdot \cdot \quad (4.)$

Regarding $(x + y)$ as a single unknown quantity (page 248),

$$x + y = -\frac{1}{2} \pm \sqrt{56 + \frac{1}{4}} = 7.$$

Substituting this value in Equation (1), we have,

$$7 + xy = 17, \quad \text{and} \quad y = 5.$$

Hence, the numbers are 2 and 5.

3. What two numbers are those whose sum is 8, and sum of their squares 34? *Ans.* 5 and 3.

4. It is required to find two such numbers, that the first shall be to the second as the second is to 16, and the sum of whose squares shall be 225?　　　　*Ans.* 9 and 12.

5. What two numbers are those which are to each other as 3 to 5, and whose squares added together make 1666?
　　　　　　　　　　　　　　　*Ans.* 21 and 35.

6. There are two numbers whose difference is 7, and half their product plus 30 is equal to the square of the less number: what are the numbers?　　　*Ans.* 12 and 19.

7. What two numbers are those whose sum is 5, and the sum of their cubes 35?　　　　　　*Ans.* 2 and 3.

8. What two numbers are those whose sum is to the greater as 11 to 7, and the difference of whose squares is 132?　　　　　　　　　　　*Ans.* 14 and 8.

9. Divide the number 100 into two such parts, that the product may be to the sum of their squares as 6 to 13.
　　　　　　　　　　　　　　*Ans.* 40 and 60.

10. Two persons, *A* and *B*, departed from different places at the same time, and traveled towards each other. On meeting, it appeared that *A* had traveled 18 miles more than *B*; and that *A* could have gone *B's* journey in $15\frac{3}{4}$ days, but *B* would have been 28 days in performing *A's* journey: how far did each travel?　　*Ans.* $\begin{cases} A, \text{ 72 miles.} \\ B, \text{ 54 miles.} \end{cases}$

11. There are two numbers whose difference is 15, and half their product is equal to the cube of the lesser number: what are those numbers?　　　　*Ans.* 3 and 18.

12. What two numbers are those whose sum, multiplied by the greater, is equal to 77; and whose difference, multiplied by the less, is equal to 12?
　　　　*Ans.* 4 and 7, or $\frac{3}{2}\sqrt{2}$ and $\frac{11}{2}\sqrt{2}$.

13. Divide 100 into two such parts, that the sum of their square roots may be 14.    *Ans.* 64 and 36.

14. It is required to divide the number 24 into two such parts, that their product may be equal to 35 times their difference.    *Ans.* 10 and 14.

15. The sum of two numbers is 8, and the sum of their cubes is 152 : what are the numbers?    *Ans.* 3 and 5.

16. Two merchants each sold the same kind of stuff; the second sold 3 yards more of it than the first, and together they receive 35 dollars. The first said to the second, "I would have received 24 dollars for your stuff;" the other replied, "And I should have received 12½ dollars for yours :" how many yards did each of them sell?

$$Ans. \begin{cases} \text{1st merchant } x' = 15, \\ \text{2d} \quad\quad\text{"} \quad\quad y' = 18, \end{cases} \text{or,} \quad \begin{matrix} x'' = 5. \\ y'' = 8. \end{matrix}$$

17. A widow possessed 13,000 dollars, which she divided into two parts, and placed them at interest in such a manner that the incomes from them were equal. If she had put out the first portion at the same rate as the second, she would have drawn for this part 360 dollars interest; and if she had placed the second out at the same rate as the first, she would have drawn for it 490 dollars interest: what were the two rates of interest?    *Ans.* 7 and 6 per cent.

18. Find three numbers, such that the difference between the third and second shall exceed the difference between the second and first by 6 ; that the sum of the numbers shall be 33, and the sum of their squares 467.

    *Ans.* 5, 9, and 19.

19. What number is that which, being divided by the product of its two digits, the quotient will be 3; and if 18 be added to it, the resulting number will be expressed by the digits inverted?    *Ans.* 24.

20. What two numbers are those which are to each other as $m$ to $n$, and the sum of whose squares is $b$ ?

$$Ans. \quad \frac{m\sqrt{b}}{\sqrt{m^2 + n^2}}, \quad \frac{n\sqrt{b}}{\sqrt{m^2 + n^2}}.$$

21. What two numbers are those which are to each other as $m$ to $n$, and the difference of whose squares is $b$ ?

$$Ans. \quad \frac{m\sqrt{b}}{\sqrt{m^2 - n^2}}, \quad \frac{n\sqrt{b}}{\sqrt{m^2 - n^2}}.$$

22. Required to find three numbers, such that the product of the first and second shall be equal to 2 ; the product of the first and third equal to 4, and the sum of the squares of the second and third equal to 20.    *Ans.* 1, 2, and 4.

23. It is required to find three numbers, whose sum shall be 38, the sum of their squares 634, and the difference between the second and first greater by 7 than the difference between the third and second.    *Ans.* 3, 15, and 20.

24. Required to find three numbers, such that the product of the first and second shall be equal to $a$ ; the product of the first and third equal to $b$ ; and the sum of the squares of the second and third equal to $c$.

$$Ans. \begin{cases} x = \sqrt{\dfrac{c}{a^2 + b^2}}. \\[2mm] y = a\sqrt{\dfrac{c}{a^2 + b^2}}. \\[2mm] z = b\sqrt{\dfrac{c}{a^2 + b^2}}. \end{cases}$$

25. What two numbers are those, whose sum, multiplied by the greater, gives 144 ; and whose difference, multiplied by the less, gives 14 ?    *Ans.* 9 and 7.

## CHAPTER IX.

### OF PROPORTIONS AND PROGRESSIONS.

**176.** Two quantities of the same kind may be compared, the one with the other, in two ways:

1st. By considering *how much* one is greater or less than the other, which is shown by their difference; and,

2d. By considering *how many times* one is greater or less than the other, which is shown by their quotient.

Thus, in comparing the numbers 3 and 12 together, with respect to their difference, we find that 12 *exceeds* 3, by 9; and in comparing them together with respect to their quotient, we find that 12 *contains* 3, four times, or that 12 is 4 times as great as 3.

The first of these methods of comparison is called *Arithmetical Proportion*, and the second, *Geometrical Proportion.*

Hence, *Arithmetical Proportion considers the relation of quantities with respect to their difference, and Geometrical Proportion the relation of quantities with respect to their quotient.*

---

176. In how many ways may two quantities be compared the one with the other? What does the first method consider? What the second? What is the first of these methods called? What is the second called? How then do you define the two proportions?

OF ARITHMETICAL PROPORTION AND PROGRESSION.

**177.** If we have four numbers, 2, 4, 8, and 10, of which the difference between the first and second is equal to the difference between the third and fourth, these numbers are said to be in arithmetical proportion. The first term 2 is called an *antecedent*, and the second term 4, with which it is compared, a *consequent*. The number 8 is also called an antecedent, and the number 10, with which it is compared, a consequent.

When the difference between the first and second is equal to the difference between the third and fourth, the four numbers are said to be in proportion. Thus, the numbers,

$$2, \quad 4, \quad 8, \quad 10,$$

are in arithmetical proportion.

**178.** When the difference between the first antecedent and consequent is the same as between any two consecutive terms of the proportion, the proportion is called an *arithmetical progression*. Hence, a *progression by differences*, or an *arithmetical progression*, is a series in which the successive terms are continually increased or decreased by a constant number, which is called the *common difference* of the progression.

Thus, in the two series,

$$1, \quad 4, \quad 7, \quad 10, \quad 13, \quad 16, \quad 19, \quad 22, \quad 25, \, \dots$$
$$60, \quad 56, \quad 52, \quad 48, \quad 44, \quad 40, \quad 36, \quad 32, \quad 28, \, \dots$$

---

177. When are four numbers in arithmetical proportion? What is the first called? What is the second called? What is the third called? What is the fourth called?

178. What is an arithmetical progression? What is the number called by which the terms are increased or diminished? What is an increasing progression? What is a decreasing progression? Which term is only an antecedent? Which only a consequent?

the first is called an *increasing progression*, of which the common difference is 3, and the second, a *decreasing progression*, of which the common difference is 4.

In general, let $a$, $b$, $c$, $d$, $e$, $f$, ... denote the terms of a progression by differences; it has been agreed to write them thus :

$$a \,.\, b \,.\, c \,.\, d \,.\, e \,.\, f \,.\, g \,.\, h \,.\, i \,.\, k \ldots$$

This series is read, $a$ is to $b$, as $b$ is to $c$, as $c$ is to $d$, as $d$ is to $e$, &c. This is a series of *continued equi-differences*, in which each term is at the same time an antecedent and a consequent, with the exception of the first term, which is only an *antecedent*, and the last, which is only a *consequent*.

**179.** Let $d$ denote the common difference of the progresion,

$$a \,.\, b \,.\, c \,.\, e \,.\, f \,.\, g \,.\, h. \text{ &c.,}$$

which we will consider increasing.

From the definition of the progression, it evidently follows that,

$$b = a + d, \quad c = b + d = a + 2d, \quad e = c + d = a + 3d;$$

and, in general, any term of the series is equal to *the first term, plus as many times the common difference as there are preceding terms.*

Thus, let $l$ be any term, and $n$ the number which marks the place of it; the expression for this *general term* is,

$$l = a + (n - 1)d.$$

Hence, for finding the last term, we have the following

---

### RULE.

I. *Multiply the common difference by the number of terms less one:*

II. *To the product add the first term; the sum will be the last term.*

### EXAMPLES.

The formula,

$$l = a + (n - 1)d,$$

serves to find any term whatever, without determining all those which precede it.

1. If we make $n = 1$, we have, $l = a$; that is, the series will have but one term.

2. If we make $n = 2$, we have, $l = a + d$; that is, the series will have two terms, and the second term is equal to the first, plus the common difference.

3. If $a = 3$, and $d = 2$, what is the 3d term?
*Ans.* 7.

4. If $a = 5$, and $d = 4$, what is the 6th term?
*Ans.* 25.

5. If $a = 7$, and $d = 5$, what is the 9th term?
*Ans.* 47.

6. If $a = 8$, and $d = 5$, what is the 10th term?
*Ans.* 53.

7. If $a = 20$, and $d = 4$, what is the 12th term?
*Ans.* 64.

8. If $a = 40$, and $d = 20$, what is the 50th term?
*Ans.* 1020.

9 If $a = 45$, and $d = 30$, what is the 40th term?
*Ans.* 1215.

10. If $a = 30$, and $d = 20$, what is the 60th term?
Ans. 1210.

11. If $a = 50$, and $d = 10$, what is the 100th term?
Ans. 1040.

12. To find the 50th term of the progression,

$$1 \,.\, 4 \,.\, 7 \,.\, 10 \,.\, 13 \,.\, 16 \,.\, 19 \ldots$$

we have, $\qquad l = 1 + 49 \times 3 = 148.$

13. To find the 60th term of the progression,

$$1 \,.\, 5 \,.\, 9 \,.\, 13 \,.\, 17 \,.\, 21 \,.\, 25 \ldots$$

we have, $\qquad l = 1 + 59 \times 4 = 237.$

**180.** If the progression were a decreasing one, we should have,

$$l = a - (n - 1)d.$$

Hence, to find the last term of a decreasing progression, we have the following

### RULE.

I. *Multiply the common difference by the number of terms less one:*

II. *Subtract the product from the first term; the remainder will be the last term.*

### EXAMPLES.

1. The first term of a decreasing progression is 60, the number of terms 20, and the common difference 3: what is the last term?

$l = a - (n - 1)d$, gives $l = 60 - (20 - 1)3 = 60 - 57 = 3.$

180. Give the rule for finding the last term of a series, when the progression is decreasing.

2. The first term is 90, the common difference 4, and the number of terms 15: what is the last term? *Ans.* 34.

3. The first term is 100, the number of terms 40, and the common difference 2: what is the last term? *Ans.* 22.

4. The first term is 80, the number of terms 10, and the common difference 4: what is the last term? *Ans.* 44.

5. The first term is 600, the number of terms 100, and the common difference 5: what is the last term?

*Ans.* 105.

6. The first term is 800, the number of terms 200, and the common difference 2: what is the last term?

*Ans.* 402.

**181.** A progression by differences being given, it is proposed to prove that, *the sum of any two terms, taken at equal distances from the two extremes, is equal to the sum of the two extremes.*

That is, if we have the progression,

$$2 . 4 . 6 . 8 . 10 . 12,$$

we wish to prove generally, that,

$$4 + 10, \quad \text{or} \quad 6 + 8,$$

is equal to the sum of the two extremes, 2 and 12.

Let $a . b . c . e . f \ldots i . k . l$, be the proposed progression, and $n$ the number of terms.

We will first observe that, if $x$ denotes a term which has $p$ terms before it, and $y$ a term which has $p$ terms after it, we have, from what has been said,

---

181. In every progression by differences, what is the sum of the two extremes equal to? What is the rule for finding the sum of an arithmetical series?

$$x = a + p \times d,$$

and,

$$y = l - p \times d;$$

whence, by addition, $x + y = a + l,$

which proves the proposition.

Referring to the previous example, if we suppose, in the first place, $x$ to denote the second term 4, then $y$ will denote the term 10, next to the last. If $x$ denotes the third term 6, then $y$ will denote 8, the third term from the last.

To apply this principle in finding the sum of the terms of a progression, write the terms, as below, and then again, in an inverse order, viz. :

$$a \,.\, b \,.\, c \,.\, d \,.\, e \,.\, f \ldots i \,.\, k \,.\, l.$$
$$l \,.\, k \,.\, i \,\ldots\ldots\ldots\, c \,.\, b \,.\, a.$$

Calling $S$ the sum of the terms of the first progression, $2S$ will be the sum of the terms of both progressions, and we shall have,

$$2S = (a+l) + (b+k) + (c+i) \ldots + (i+c) + (k+b) + (l+a).$$

Now, since all the parts, $a + l$, $b + k$, $c + i \ldots$ are equal to each other, and their number equal to $n$,

$$2S = (a + l) \times n, \quad \text{or} \quad S = \left(\frac{a + l}{2}\right) \times n.$$

Hence, for finding the sum of an arithmetical series, we have the following

### RULE.

I. *Add the two extremes together, and take half their sum:*

II. *Multiply this half-sum by the number of terms; the product will be the sum of the series.*

12

EXAMPLES.

1. The extremes are 2 and 16, and the number of terms 8 : what is the sum of the series?

$$S = \left(\frac{a+l}{2}\right) \times n, \quad \text{gives} \quad S = \frac{2+16}{2} \times 8 = 72.$$

2. The extremes are 3 and 27, and the number of terms 12 : what is the sum of the series?          *Ans.* 180.

3. The extremes are 4 and 20, and the number of terms 10 : what is the sum of the series?          *Ans.* 120.

4. The extremes are 100 and 200, and the number of terms 80 : what is the sum of the series?          *Ans.* 12000.

5. The extremes are 500 and 60, and the number of terms 20 : what is the sum of the series?          *Ans.* 5600

6. The extremes are 800 and 1200, and the number of terms 50 : what is the sum of the series?          *Ans.* 50000.

**182.** In arithmetical proportion there are five members to be considered :

1st.  The first term, $a$.
2d.  The common difference, $d$.
3d.  The number of terms, $n$.
4th.  The last term, $l$.
5th.  The sum, $S$.

The formulas,

$$l = a + (n-1)d, \quad \text{and} \quad S = \left(\frac{a+l}{2}\right) \times n,$$

contain five quantities, $a$, $d$, $n$, $l$, and $S$, and consequently give rise to the following general problem, viz.: *Any three*

---

182. How many numbers are considered in arithmetical proportion? What are they? In every arithmetical progression, what is the common difference equal to?

*of these five quantities being given, to determine the other two.*

We already know the value of $S$ in terms of $a$, $n$, and $l$. From the formula,

$$l = a + (n - 1)d,$$

we find, $\qquad a = l - (n - 1)d.$

That is: *The first term of an increasing arithmetical progression is equal to the last term, minus the product of the common difference by the number of terms less one.*

From the same formula, we also find,

$$d = \frac{l - a}{n - 1}.$$

That is: *In any arithmetical progression, the common difference is equal to the last term, minus the first term, divided by the number of terms less one.*

The last term is 16, the first term 4, and the number of terms 5: what is the common difference?

The formula, $\qquad d = \dfrac{l - a}{n - 1}$

gives, $\qquad d = \dfrac{16 - 4}{4} = 3.$

2. The last term is 22, the first term 4, and the number of terms 10: what is the common difference? *Ans.* 2.

**183.** The last principle affords a solution to the following question:

*To find a number* m *of arithmetical means between two given numbers* a *and* b.

---

183. How do you find any number of arithmetical means between two given numbers?

To resolve this question, it is first necessary to find the common difference. Now, we may regard $a$ as the first term of an arithmetical progression, $b$ as the last term, and the required means as intermediate terms. The number of terms of this progression will be expressed by $m + 2$.

Now, by substituting in the above formula, $b$ for $l$, and $m + 2$ for $n$, it becomes,

$$d = \frac{b - a}{m + 2 - 1} = \frac{b - a}{m + 1};$$

that is: *The common difference of the required progression is obtained by dividing the difference between the given numbers, a and b, by the required number of means plus one.*

Having obtained the common difference, $d$, form the second term of the progression, or the *first arithmetical mean,* by adding $d$ to the first term $a$. The *second mean* is obtained by augmenting the first mean by $d$, &c.

1. Find three arithmetical means between the extremes 2 and 18.

The formula, $$d = \frac{b - a}{m + 1},$$

gives, $$d = \frac{18 - 2}{4} = 4;$$

hence, the progression is,

$$2 \;.\; 6 \;.\; 10 \;.\; 14 \;.\; 18.$$

2. Find twelve arithmetical means between 12 and 77.

The formula, $$d = \frac{b - a}{m + 1},$$

gives, $$d = \frac{77 - 12}{13} = 5;$$

hence, the progression is,

$$12 \;.\; 17 \;.\; 22 \;.\; 27 \;\ldots\; 77.$$

**184.** REMARK.—If the same number of arithmetical means are inserted between all the terms, taken two and two, these terms, and the arithmetical means united, will form one and the same progression.

For, let $a \cdot b \cdot c \cdot e \cdot f \ldots$ be the proposed progression, and $m$ the number of means to be inserted between $a$ and $b$, $b$ and $c$, $c$ and $e \ldots$ &c.

From what has just been said, the common difference of each partial progression will be expressed by

$$\frac{b-a}{m+1}, \quad \frac{c-b}{m+1}, \quad \frac{e-c}{m+1} \ldots$$

expressions which are equal to each other, since $a, b, c \ldots$ are in progression; therefore, the common difference is the same in each of the partial progressions; and, since the *last term* of the first forms the *first term* of the second, &c., we may conclude, that all of these partial progressions form a single progression.

### EXAMPLES.

1. Find the sum of the first fifty terms of the progression $2 \cdot 9 \cdot 16 \cdot 23 \ldots$

For the 50th term, we have,

$$l = 2 + 49 \times 7 = 345.$$

Hence, $\quad S = (2 + 345) \times \dfrac{50}{2} = 347 \times 25 = 8675.$

2. Find the 100th term of the series $2 \cdot 9 \cdot 16 \cdot 23 \ldots$
*Ans.* 695.

3. Find the sum of 100 terms of the series $1 \cdot 3 \cdot 5 \cdot 7 \cdot 9 \ldots$
*Ans.* 10000.

4. The greatest term is 70, the common difference 3, and the number of terms 21 : what is the least term and the sum of the series?

Ans. Least term, 10 ; sum of series, 840.

5. The first term is 4, the common difference 8, and the number of terms 8 : what is the last term, and the sum of the series? Ans. $\begin{cases} \text{Last term, } 60. \\ \text{Sum } = 256. \end{cases}$

6. The first term is 2, the last term 20, and the number of terms 10 : what is the common difference? Ans. 2.

7. Insert four means between the two numbers 4 and 19 : what is the series? Ans. 4 . 7 . 10 . 13 . 16 . 19.

8. The first term of a decreasing arithmetical progression is 10, the common difference one-third, and the number of terms 21 : required the sum of the series. Ans. 140.

9. In a progression by differences, having given the common difference 6, the last term 185, and the sum of the terms 2945 : find the first term, and the number of terms.

Ans. First term = 5 ; number of terms, 31.

10. Find nine arithmetical means between each antecedent and consequent of the progression 2 . 5 . 8 . 11 . 14 ...
Ans. Common diff., or $d = 0.3$.

11. Find the number of men contained in a triangular battalion, the first rank containing one man, the second 2, the third 3, and so on to the $n^{th}$, which contains $n$. In other words, find the expression for the sum of the natural numbers 1, 2, 3 ..., from 1 to $n$ inclusively.

$$Ans. \quad S = \frac{n(n+1)}{2}.$$

12. Find the sum of the $n$ first terms of the progression of uneven numbers, 1 . 3 . 5 . 7 . 9, ... Ans. $S = n^2$.

13. One hundred stones being placed on the ground in a straight line, at the distance of 2 yards apart, how far will a person travel who shall bring'them one by one to a basket, placed at a distance of 2 yards from the first stone?

*Ans.* 11 miles, 840 yards.

## GEOMETRICAL PROPORTION AND PROGRESSION.

**185.** *Ratio* is the quotient arising from dividing one quantity by another quantity of the same kind, regarded as a standard. Thus, if the numbers 3 and 6 have the same unit, the ratio of 3 to 6 will be expressed by

$$\frac{6}{3} = 2.$$

And in general, if $A$ and $B$ represent quantities of the same kind, the ratio of $A$ to $B$ will be expressed by

$$\frac{B}{A}.$$

**186.** The character $\propto$ indicates that one quantity is proportional to another. Thus,

$$A \propto B,$$

is read, $A$ proportional to $B$.

If there be four numbers,

$$2, \quad 4, \quad 8, \quad 16,$$

having such values that the second divided by the first is equal to the fourth divided by the third, the numbers are

185. What is ratio? What is the ratio of 3 to 6? Of 4 to 12?

186. What is proportion? How do you express that four numbers are in proportion? What are the numbers called? What are the first and fourth terms called? What the second and third?

said to form a proportion.  And in general, if there be four quantities, $A$, $B$, $C$, and $D$, having such values that,

$$\frac{B}{A} = \frac{D}{C},$$

then, $A$ is said to have the same ratio to $B$ that $C$ has to $D$; or, the ratio of $A$ to $B$ is equal to the ratio of $C$ to $D$. When four quantities have this relation to each other, compared together two and two, they are said to form a geometrical proportion.

To express that the ratio of $A$ to $B$ is equal to the ratio of $C$ to $D$, we write the quantities thus,

$$A : B :: C : D;$$

and read, $A$ is to $B$ as $C$ to $D$.

The quantities which are compared, the one with the other, are called *terms* of the proportion.  The first and last terms are called the *two extremes*, and the second and third terms, the *two means*.  Thus, $A$ and $D$ are the extremes, and $B$ and $C$ the means.

**187.**  Of four terms of a proportion, the first and third are called the *antecedents*, and the second and fourth the *consequents ;* and the last is said to be a fourth proportional to the other three, taken in order.  Thus, in the last proportion $A$ and $C$ are the antecedents, and $B$ and $D$ the consequents.

**188.**  Three quantities are in proportion, when the first has the same ratio to the second that the second has to the

---

187. In four proportional quantities, what are the first and third called? What the second and fourth ?

188. When are three quantities proportional?  What is the middle one called?

third; and then the middle term is said to be a mean proportional between the other two. For example,

$$3 : 6 :: 6 : 12;$$

and 6 is a mean proportional between 3 and 12. –

**189.** Four quantities are said to be in proportion by *inversion*, or *inversely*, when the consequents are made the antecedents, and the antecedents the consequents.

Thus, if we have the proportion,

$$3 : 6 :: 8 : 16,$$

the inverse proportion would be,

$$6 : 3 :: 16 : 8.$$

**190.** Quantities are said to be in proportion by *alternation*, or *alternately*, when antecedent is compared with antecedent, and consequent with consequent.

Thus, if we have the proportion,

$$3 : 6 :: 8 : 16,$$

the alternate proportion would be,

$$3 : 8 :: 6 : 16.$$

**191.** Quantities are said to be in proportion by *composition*, when the sum of the antecedent and consequent is compared either with antecedent or consequent

Thus, if we have the proportion,

$$2 : 4 :: 8 : 16,$$

189. When are quantities said to be in proportion by inversion, or inversely?
190. When are quantities in proportion by alternation?
191. When are quantities in proportion by composition?
12*

# 274    ELEMENTARY ALGEBRA.

the proportion by composition would be,

$$2 + 4 : 2 :: 8 + 16 : 8;$$

and, $\quad\quad 2 + 4 : 4 :: 8 + 16 : 16.$

**192.** Quantities are said to be in proportion by *division*, when the difference of the antecedent and consequent is compared either with antecedent or consequent.

Thus, if we have the proportion,

$$3 : 9 :: 12 : 36,$$

the proportion by division will be,

$$9 - 3 : 3 :: 36 - 12 : 12;$$

and, $\quad\quad 9 - 3 : 9 :: 36 - 12 : 36;$

**193.** Equi-multiples of two or more quantities are the products which arise from multiplying the quantities by the same number.

Thus, if we have any two numbers, as 6 and 5, and multiply them both by any number, as 9, the equi-multiples will be 54 and 45; for,

$$6 \times 9 = 54, \quad \text{and} \quad 5 \times 9 = 45.$$

Also, $m \times A$, and $m \times B$, are equi-multiples of $A$ and $B$, the common multiplier being $m$.

**194.** Two quantities $A$ and $B$, which may change their values, *are reciprocally or inversely proportional, when one is proportional to unity divided by the other, and then their product remains constant.*

---

192. When are quantities in proportion by division ?
193. What are equi-multiples of two or more quantities ?
194. When are two quantities said to be reciprocally proportional ?

We express this reciprocal or inverse relation thus,

$$A \propto \frac{1}{B}.$$

in which $A$ is said to be inversely proportional to $B$.

**195.** If we have the proportion,

$$A : B :: C : D,$$

we have,    $$\frac{B}{A} = \frac{D}{C}, \quad \text{(Art. 186)};$$

and by clearing the equation of fractions, we have,

$$BC = AD.$$

That is: *Of four proportional quantities, the product of the two extremes is equal to the product of the two means.*

This general principle is apparent in the proportion between the numbers,

$$2 : 10 :: 12 : 60,$$

which gives,    $2 \times 60 = 10 \times 12 = 120.$

**196.** If four quantities, $A, B, C, D$, are so related to each other, that

$$A \times D = B \times C,$$

we shall also have,    $$\frac{B}{A} = \frac{D}{C};$$

and hence,    $$A : B :: C : D.$$

That is: *If the product of two quantities is equal to the product of two other quantities, two of them may be made the extremes, and the other two the means of a proportion.*

---

195. If four quantities are proportional, what is the product of the two means equal to?

196. If the product of two quantities is equal to the product of two other quantities, may the four be placed in a proportion? How?

Thus, if we have,

$$2 \times 8 = 4 \times 4,$$

we also have,

$$2 : 4 :: 4 : 8.$$

**197.**  If we have three proportional quantities,

$$A : B :: B : C,$$

we have,

$$\frac{B}{A} = \frac{C}{B};$$

hence,

$$B^2 = AC.$$

That is: *If three quantities are proportional, the square of the middle term is equal to the product of the two extremes.*

Thus, if we have the proportion,

$$3 : 6 :: 6 : 12,$$

we shall also have,

$$6 \times 6 = 6^2 = 3 \times 12 = 36.$$

**198.**  If we have,

$$A : B :: C : D, \text{ and consequently, } \frac{B}{A} = \frac{D}{C},$$

multiply both members of the last equation by $\frac{C}{B}$, and we then obtain,

$$\frac{C}{A} = \frac{D}{B};$$

and, hence,     $A : C :: B : D.$

That is: *If four quantities are proportional, they will be in proportion by alternation.*

---

197. If three quantities are proportional, what is the product of the extremes equal to?

198. If four quantities are proportional, will they be in proportion by alternation?

Let us take, as an example,

$$10 : 15 :: 20 : 30.$$

We shall have, by alternating the terms,

$$10 : 20 :: 15 : 30.$$

**199.** If we have,

$$A : B :: C : D, \text{ and } A : B :: E : F,$$

we shall also have,

$$\frac{B}{A} = \frac{D}{C}, \text{ and } \frac{B}{A} = \frac{F}{E};$$

hence, $\quad \dfrac{D}{C} = \dfrac{F}{E}, \text{ and } C : D :: E : F.$

That is: *If there are two sets of proportions having an antecedent and consequent in the one, equal to an antecedent and consequent of the other, the remaining terms will be proportional.*

If we have the two proportions,

$$2 : 6 :: 8 : 24, \text{ and } 2 : 6 :: 10 : 30,$$

we shall also have,

$$8 : 24 :: 10 : 30.$$

**200.** If we have,

$$A : B :: C : D, \text{ and consequently, } \frac{B}{A} = \frac{D}{C},$$

we have, by dividing 1 by each member of the equation,

$$\frac{A}{B} = \frac{C}{D}, \text{ and consequently, } B : A :: D : C.$$

---

199. If you have two sets of proportions having an antecedent and consequent in each, equal ; what will follow ?

200. If four quantities are in proportion, will they be in proportion when taken inversely ?

That is: *Four proportional quantities will be in proportion, when taken inversely.*

To give an example in numbers, take the proportion,

$$7 : 14 :: 8 : 16;$$

then, the inverse proportion will be,

$$14 . 7 :: 16 : 8,$$

in which the ratio is one-half.

**201.**   The proportion,

$$A : B :: C : D, \text{ gives, } A \times D = B \times C.$$

To each member of the last equation add $B \times D$. We shall then have,

$$(A + B) \times D = (C + D) \times B;$$

and by separating the factors, we obtain,

$$A + B : B :: C + D : D.$$

If, instead of adding, we subtract $B \times D$ from both members, we have,

$$(A - B) \times D = (C - D) \times B;$$

which gives,

$$A - B : B :: C - D : D.$$

That is: *If four quantities are proportional, they will be in proportion by composition or division.*

Thus, if we have the proportion,

$$9 : 27 :: 16 : 48,$$

---

201. If four quantities are in proportion, will they be in proportion by composition? Will they be in proportion by division? What is the difference between composition and division?

we shall have, by composition,

$$9 + 27 : 27 :: 16 + 48 : 48;$$

that is, $$36 : 27 :: 64 : 48,$$

in which the ratio is three-fourths.

The same proportion gives us, by division,

$$27 - 9 : 27 :: 48 - 16 : 48;$$

that is, $$18 : 27 :: 32 : 48,$$

in which the ratio is one and one-half.

**202.** If we have,

$$\frac{B}{A} = \frac{D}{C},$$

and multiply the numerator and denominator of the first member by any number $m$, we obtain,

$$\frac{mB}{mA} = \frac{D}{C}, \text{ and } mA : mB :: C : D.$$

That is: *Equal multiples of two quantities have the same ratio as the quantities themselves.*

For example, if we have the proportion,

$$5 : 10 :: 12 : 24,$$

and multiply the first antecedent and consequent by 6, we have,

$$30 : 60 :: 12 : 24,$$

in which the ratio is still 2.

**203.** The proportions,

$$A : B :: C : D, \text{ and } A : B :: E : F,$$

---

202. Have equal multiples of two quantities the same ratio as the quantities?

203 Suppose the antecedent and consequent be augmented or diminished by quantities having the same ratio?

give, $A \times D = B \times C$, and $A \times F = B \times E$;

adding and subtracting these equations, we obtain,

$$A(D \pm F) = B(C \pm E), \quad \text{or} \quad A : B :: C \pm E : D \pm F.$$

That is: *If C and D, the antecedent and consequent, be augmented or diminished by quantities E and F, which have the same ratio as C to D, the resulting quantities will also have the same ratio.*

Let us take, as an example, the proportion,

$$9 : 18 :: 20 : 40,$$

in which the ratio is 2.

If we augment the antecedent and consequent by the numbers 15 and 30, which have the same ratio, we shall have,

$$9 + 15 : 18 + 30 :: 20 : 40;$$

that is,          $24 : 48 :: 20 : 40,$

in which the ratio is still 2.

If we diminish the second antecedent and consequent by these numbers respectively, we have,

$$9 : 18 :: 20 - 15 : 40 - 30;$$

that is,          $9 : 18 :: 5 : 10,$

in which the ratio is till 2.

**204.**  If we have several proportions,

$A : B :: C : D,$  which gives  $A \times D = B \times C,$

$A : B :: E : F,$  which gives  $A \times F = B \times E,$

$A : B :: G : H,$  which gives  $A \times H = B \times G,$

&c., &c.,

---

204. In any number of proportions having the same ratio, how will any one antecedent be to its consequent?

we shall have, by addition,

$$A(D + F + H) = B(C + E + G);$$

and by separating the factors,

$$A : B :: C + E + G : D + F + H.$$

That is: *In any number of proportions having the same ratio, any antecedent will be to its consequent as the sum of the antecedents to the sum of the consequents.*

Let us take, for example,

$$2 : 4 :: 6 : 12, \quad \text{and} \quad 1 : 2 :: 3 : 6, \quad \&c.$$

Then $$2 : 4 :: 6 + 3 : 12 + 6;$$

that is, $$2 : 4 :: 9 : 18,$$

in which the ratio is still 2.

**205.** If we have four proportional quantities,

$$A : B :: C : D, \quad \text{we have,} \quad \frac{B}{A} = \frac{D}{C};$$

and raising both members to any power whose exponent is $n$, or extracting any root whose index is $n$, we have,

$$\frac{B^n}{A^n} = \frac{D^n}{C^n}, \quad \text{and consequently,}$$

$$A^n : B^n :: C^n : D^n.$$

That is: *If four quantities are proportional, their like powers or roots will be proportional.*

If we have, for example,

$$2 : 4 :: 3 : 6,$$

we shall have, $$2^2 : 4^2 :: 3^2 : 6^2;$$

---

205. In four proportional quantities, how are like powers or roots?

that is,                  4 : 16 :: 9 : 36,

in which the terms are proportional, the ratio being 4.

**206.** Let there be two sets of proportions,

$$A : B :: C : D, \quad \text{which gives} \quad \frac{B}{A} = \frac{D}{C};$$

$$E : F :: G : H, \quad \text{which gives} \quad \frac{F}{E} = \frac{H}{G}.$$

Multiply them together, member by member, we have,

$$\frac{B \times F}{A \times E} = \frac{D \times H}{C \times G},$$

$$A \times E : B \times F :: C \times G : D \times H.$$

That is : *In two sets of proportional quantities, the products of the corresponding terms are proportional.*

Thus, if we have the two proportions,

8 : 16 :: 10 :  20,

and,              3 :  4 ::  6 :   8,

we shall have,    24 : 64 :: 60 : 160.

———

GEOMETRICAL PROGRESSION.

**207.** We have thus far only considered the case in which the ratio of the first term to the second is the same as that of the third to the fourth.

———

206. In two sets of proportions, how are the products of the corresponding terms?

207. What is a geometrical progression? What is the ratio of the progression? If any term of a progression be multiplied by the ratio, what will the product be? If any term be divided by the ratio, what

If we have the farther condition, that the ratio of the second term to the third shall also be the same as that of the first to the second, or of the third to the fourth, we shall have a series of numbers, each one of which, divided by the preceding one, will give the same ratio. Hence, if any term be multiplied by this quotient, the product will be the succeeding term. A series of numbers so formed, is called a *geometrical progression*. Hence,

A *Geometrical Progression*, or *progression by quotients*, is a series of terms, each of which is equal to the preceding term multiplied by a *constant number*, which number is called the *ratio* of the progression. Thus,

$$1 : 3 : 9 : 27 : 81 : 243, \text{ &c.,}$$

is a geometrical progression, in which the ratio is 3. It is written by merely placing two dots between the terms.

Also,   $64 : 32 : 16 : 8 : 4 : 2 : 1,$

is a geometrical progression in which the ratio is *one-half*.

In the first progression each term is contained three times in the one that follows, and hence the ratio is 3. In the second, each term is contained one-half times in the one which follows, and hence the ratio is one-half.

The first is called an *increasing* progression, and the second a *decreasing* progression.

Let $a$, $b$, $c$, $d$, $e$, $f$, . . . be numbers, in a progression by quotients; they are written thus:

$$a : b : c : d : e : f : g \ldots$$

and it is enunciated in the same manner as a progression by differences. It is necessary, however, to make the distinc-

will the quotient be? How is a progression by quotients written? Which of the terms is only an antecedent? Which only a consequent? How may each of the others be considered?

tion, that one is a series formed by equal differences, and the other a series formed by equal quotients or ratios. It should be remarked that each term is at the same time an antecedent and a consequent, except the first, which is only an antecedent, and the last, which is only a consequent.

**208.** Let $r$ denote the ratio of the progression,

$$a : b : c : d \ldots$$

$r$ being $> 1$ when the progression is *increasing*, and $r < 1$ when it is *decreasing*. Then, since,

$$\frac{b}{a} = r, \quad \frac{c}{b} = r, \quad \frac{d}{c} = r, \quad \frac{e}{d} = r, \quad \&c.,$$

we have,

$$b = ar, \quad c = br = ar^2, \quad d = cr = ar^3, \quad e = dr = ar^4,$$
$$f = er = ar^5 \ldots$$

that is, the second term is equal to $ar$, the third to $ar^2$, the fourth to $ar^3$, the fifth to $ar^4$, &c.; and in general, the $n$th term, that is, one which has $n - 1$ terms before it, is expressed by $ar^{n-1}$.

Let $l$ be this term; we then have the formula,

$$l = ar^{n-1},$$

by means of which we can obtain any term without being obliged to find all the terms which precede it. Hence, to find the last term of a progression, we have the following

RULE.

I. *Raise the ratio to a power whose exponent is one less than the number of terms.*

II. *Multiply the power thus found by the first term: the product will be the required term.*

---

208. By what letter do we denote the ratio of a progression? In an increasing progression is $r$ greater or less than 1 ? In a decreasing pro-

## EXAMPLES.

1. Find the 5th term of the progression,

$$2 : 4 : 8 : 16 \ldots$$

in which the first term is 2, and the common ratio 2.

5th term $= 2 \times 2^4 = 2 \times 16 = 32.$ *Ans.*

2. Find the 8th term of the progression,

$$2 : 6 : 18 : 54 \ldots$$

8th term $= 2 \times 3^7 = 2 \times 2187 = 4374.$ *Ans.*

3. Find the 6th term of the progression,

$$2 : 8 : 32 : 128 \ldots$$

6th term $= 2 \times 4^5 = 2 \times 1024 = 2048.$ *Ans*

4. Find the 7th term of the progression,

$$3 : 9 : 27 : 81 \ldots$$

7th term $= 3 \times 3^6 = 3 \times 729 = 2187.$ *Ans.*

5. Find the 6th term of the progression,

$$4 : 12 : 36 : 108 \ldots$$

6th term $= 4 \times 3^5 = 4 \times 243 = 972.$ *Ans.*

6. A person agreed to pay his servant 1 cent for the first day, two for the second, and four for the third, doubling every day for ten days: how much did he receive on the tenth day? *Ans.* $5.12.

gression is $r$ greater or less than 1? If $a$ is the first term and $r$ the ratio, what is the second term equal to? What the third? What the fourth? What is the last term equal to? Give the rule for finding the last term.

7. What is the 8th term of the progression,

$$9 : 36 : 144 : 576 \ldots$$

8th term $= 9 \times 4^7 = 9 \times 16384 = 147456$. *Ans.*

8. Find the 12th term of the progression,

$$64 : 16 : 4 : 1 : \frac{1}{4} \ldots$$

$$12\text{th term} = 64\left(\frac{1}{4}\right)^{11} = \frac{4^3}{4^{11}} = \frac{1}{4^8} = \frac{1}{65536}. \ Ans.$$

**209.** We will now proceed to determine the sum of $n$ terms of a progression,

$$a : b : c : d : e : f : \ldots : i : k : l;$$

$l$ denoting the $n$th term.

We have the equations (Art. 208),

$$b = ar, \quad c = br, \quad d = cr, \quad e = dr, \ldots k = ir, \quad l = kr,$$

and by adding them all together, member to member, we deduce,

*Sum of 1st members.*        *Sum of 2d members.*

$$b + c + d + e + \ldots + k + l = (a + b + c + d + \ldots + i + k)r;$$

in which we see that the first member contains all the terms but $a$, and the polynomial, within the parenthesis in the second member, contains all the terms but $l$. Hence, if we call the sum of the terms $S$, we have,

$$S - a = (S - l)r = Sr - lr, \quad \therefore \ Sr - S = lr - a;$$

whence, $\qquad\qquad S = \dfrac{lr - a}{r - 1}.$

---

209. Give the rule for finding the sum of the series. What is the first step? What the second? What the third?

Therefore, to obtain the sum of all the terms, or sum of the series of a geometrical progression, we have the

RULE.

I. *Multiply the last term by the ratio :*

II. *Subtract the first term from the product :*

III. *Divide the remainder by the ratio diminished* by 1 *and the quotient will be the sum of the series.*

1. Find the sum of eight terms of the progression,

$$2 : 6 : 18 : 54 : 162 \ldots 2 \times 3^7 = 4374.$$

$$S = \frac{lr - a}{r - 1} = \frac{13122 - 2}{2} = 6560.$$

2. Find the sum of the progression,

$$2 : 4 : 8 : 16 : 32.$$

$$S = \frac{lr - a}{r - 1} = \frac{64 - 2}{1} = 62.$$

3. Find the sum of ten terms of the progression,

$$2 : 6 : 18 : 54 : 162 \ldots 2 \times 3^9 = 39366.$$

*Ans.* 59048.

4. What debt may be discharged in a year, or twelve months, by paying $1 the first month, $2 the second month, $4 the third month, and so on, 'each succeeding payment being double the last; and what will be the last payment?

*Ans.* { Debt, . . $4095.<br>{ Last payment, $2048.

5. A daughter was married on New-Year's day. Her father gave her 1s., with an agreement to double it on the first of the next month, and at the beginning of each succeeding month to double what she had previously received. How much did she receive?     *Ans.* £204 15s.

6. A man bought ten bushels of wheat, on the condition that he should pay 1 cent for the first bushel, 3 for the second, 9 for the third, and so on to the last: what did he pay for the last bushel, and for the ten bushels?

$$Ans. \begin{cases} \text{Last bushel, } \$196\ 83. \\ \text{Total cost, } \quad \$295\ 24. \end{cases}$$

7. A man plants 4 bushels of barley, which, at the first harvest, produced 32 bushels; these he also plants, which, in like manner, produce 8 fold; he again plants all his crop, and again gets 8 fold, and so on for 16 years: what is his last crop, and what the sum of the series?

$$Ans. \begin{cases} \text{Last, } 140737488355328 \text{ bush.} \\ \text{Sum, } 160842843834660. \end{cases}$$

**210.** When the progression is decreasing, we have, $r < 1$, and $l < a$; the above formula,

$$S = \frac{lr - a}{r - 1},$$

for the sum, is then written under the form,

$$S = \frac{a - lr}{1 - r},$$

in order that the two terms of the fraction may be positive.

1. Find the sum of the terms of the progression,

$$32 : 16 : 8 : 4 : 2$$

$$S = \frac{a - lr}{1 - r} = \frac{32 - 2 \times \frac{1}{2}}{\frac{1}{2}} = \frac{31}{\frac{1}{2}} = 62.$$

---

210. What is the formula for the sum of the series of a decreasing progression?

2. Find the sum of the first twelve terms of the progression,

$$64 : 16 : 4 : 1 : \frac{1}{4} : \ldots : 64\left(\frac{1}{4}\right)^{11}, \quad \text{or} \quad \frac{1}{65536}.$$

$$S = \frac{a - lr}{1 - r} = \frac{64 - \dfrac{1}{65536} \times \dfrac{1}{4}}{\dfrac{3}{4}} = \frac{256 - \dfrac{1}{65536}}{3} = .85 + \frac{65535}{196608}$$

**211.** REMARK.—We perceive that the principal difficulty consists in obtaining the numerical value of the last term, a tedious operation, even when the number of terms is not very great.

3. Find the sum of six terms of the progression,

$$512 : 128 : 32 \ldots$$

*Ans.* 682¼.

4. Find the sum of seven terms of the progression,

$$2187 : 729 : 243 \ldots$$

*Ans.* 3279.

·5. Find the sum of six terms of the progression,

$$972 : 324 : 108 \ldots$$

*Ans.* 1456.

6. Find the sum of eight terms of the progression,

$$147456 : 36864 : 9216 \ldots$$

*Ans.* 196605.

OF PROGRESSIONS HAVING AN INFINITE NUMBER OF TERMS.

**212.** Let there be the decreasing progression,

$$a : b : c : d : e : f : \ldots$$

---

212. When the progression is decreasing, and the number of terms infinite, what is the expression for the value of the sum of the series?

13

containing an indefinite number of terms.   In the formula,

$$S = \frac{a - lr}{1 - r},$$

substitute for $l$ its value, $ar^{n-1}$, (Art. 208), and we have,

$$S = \frac{a - ar^n}{1 - r},$$

which expresses the sum of $n$ terms of the progression. This may be put under the form,

$$S = \frac{a}{1 - r} - \frac{ar^n}{1 - r}.$$

Now, since the progression is decreasing, $r$ is a proper fraction; and $r^n$ is also a fraction, which diminishes as $n$ increases.   Therefore, the greater the number of terms we take, the more will $\frac{a}{1 - r} \times r^n$ diminish, and consequently, the more will the entire sum of all the terms approximate to an equality with the first part of $S$, that is, to $\frac{a}{1 - r}$. Finally, when $n$ is taken greater than any given number, or $n =$ infinity, then $\frac{a}{1 - r} \times r^n$ will be less than any given number, or will become equal to $0$; and the expression, $\frac{a}{1 - r}$, will then represent the true value of the sum of all the terms of the series.   Whence we may conclude, that the expression for *the sum of the terms of a decreasing progression, in which the number of terms is infinite, is,*

$$S = \frac{a}{1 - r};$$

that is, *equal to the first term, divided by* 1 *minus the ratio.*

This is, properly speaking, the *limit* to which the *partial sums* approach, as we take a greater number of terms in the progression. The difference between these sums and $\dfrac{a}{1-r}$, may be made as small as we please, but will only become *nothing* when the number of terms is infinite.

### EXAMPLES.

1. Find the sum of

$$1 : \frac{1}{3} : \frac{1}{9} : \frac{1}{27} : \frac{1}{81}, \text{ to infinity.}$$

We have, for the expression of the sum of the terms,

$$S = \frac{a}{1-r} = \frac{1}{1-\dfrac{1}{3}} = \frac{3}{2} = 1\tfrac{1}{2}. \quad Ans.$$

The error committed by taking this expression for the value of the sum of the $n$ first terms, is expressed by

$$\frac{a}{1-r} \times r^n = \frac{3}{2}\left(\frac{1}{3}\right)^n.$$

First take $n = 5$; it becomes,

$$\frac{3}{2}\left(\frac{1}{3}\right)^5 = \frac{1}{2 \cdot 3^4} = \frac{1}{162}.$$

When $n = 6$, we find,

$$\frac{3}{2}\left(\frac{1}{3}\right)^6 = \frac{1}{162} \times \frac{1}{3} = \frac{1}{486}.$$

Hence, we see, that the *error committed* by taking $\dfrac{3}{2}$ for the sum of a certain number of terms, is less in proportion as this number is greater.

2. Again, take the progression,

$$1 : \frac{1}{2} : \frac{1}{4} : \frac{1}{8} : \frac{1}{16} : \frac{1}{32} : \&c. \ldots$$

We have, $S = \dfrac{a}{1 - r} = \dfrac{1}{1 - \dfrac{1}{2}} = 2.$ *Ans.*

3. What is the sum of the progression,

$$1, \quad \frac{1}{10}, \quad \frac{1}{100}, \quad \frac{1}{1000}, \quad \frac{1}{10000}, \quad \&c., \text{ to infinity.}$$

$$S = \frac{a}{1 - r} = \frac{1}{1 - \dfrac{1}{10}} = 1\frac{1}{9}. \quad \textit{Ans.}$$

**213.** In the several questions of geometrical progression, there are five numbers to be considered:

1st. The first term,   .   .   $a$.
2d. The ratio,   .   .   .   .   $r$.
3d. The number of terms, $n$.
4th. The last term,   .   .   $l$.
5th. The sum of the terms, $S$.

**214.** We shall terminate this subject by solving this problem:

To find a mean proportional between any two numbers, as $m$ and $n$.

Denote the required mean by $x$. We shall then have (Art. 197),

$$x^2 = m \times n;$$

and hence,  $\quad x = \overline{\sqrt{m \times n}}.$

213. How many numbers are considered in a geometrical progression? What are they?

214. How do you find a mean proportional between two numbers?

That is: *Multiply the two numbers together, and extract the square root of the product.*

1. What is the geometrical mean between the numbers 2 and 8?

$$\text{Mean} = \sqrt{8 \times 2} = \sqrt{16} = 4. \quad Ans.$$

2. What is the mean between 4 and 16?          *Ans.* 8.

3. What is the mean between 3 and 27?          *Ans.* 9.

4. What is the mean between 2 and 72?          *Ans.* 12.

5. What is the mean between 4 and 64?          *Ans.* 16.

therefore, $40 satisfies the enunciation.

## CHAPTER X.

### OF LOGARITHMS.

**215.** THE nature and properties of the logarithms in common use, will be readily understood by considering attentively the different powers of the number 10. They are,

$$10^0 = 1$$
$$10^1 = 10$$
$$10^2 = 100$$
$$10^3 = 1000$$
$$10^4 = 10000$$
$$10^5 = 100000$$
$$\&c., \quad \&c.$$

It is plain that the *exponents* 0, 1, 2, 3, 4, 5, &c., form an arithmetical series of which the common difference is 1; and that the numbers 1, 10, 100, 1000, 10000, 100000, &c., form a geometrical progression of which the common ratio is 10. The number 10 is called the *base* of the system of logarithms; and the exponents 0, 1, 2, 3, 4, 5, &c., are the *logarithms* of

215. What relation exists between the exponents 1, 2, 3, &c.? How are the corresponding numbers 10, 100, 1000? What is the common difference of the exponents? What is the common ratio of the corresponding numbers? What is the base of the common system of logarithms? What are the exponents? Of what number is the exponent 1 the logarithm? The exponent 2? The exponent 3?

the numbers which are produced by raising 10 to the powers denoted by those exponents.

**216.** If we denote the logarithm of any number by $m$, then the number itself will be the $m$th power of 10; that is, if we represent the corresponding number by $M$,

$$10^m = M.$$

Thus, if we make $m = 0$, $M$ will be equal to 1; if $m = 1$, $M$ will be equal to 10, &c. Hence,

*The logarithm of a number is the exponent of the power to which it is necessary to raise the base of the system in order to produce the number.*

**217.** If, as before, 10 denotes the base of the system of logarithms, $m$ any exponent, and $M$ the corresponding number, we shall then have,

$$10^m = M, \qquad (1.)$$

in which $m$ is the logarithm of $M$.

If we take a second exponent $n$, and let $N$ denote the corresponding number, we shall have,

$$10^n = N, \qquad (2.)$$

in which $n$ is the logarithm of $N$.

If, now, we multiply the first of these equations by the second, member by member, we have,

$$10^m \times 10^n = 10^{m+n} = M \times N;$$

but since 10 is the base of the system, $m + n$ is the logarithm $M \times N$; hence,

---

216. If we denote the base of a system by 10, and the exponent by $m$, what will represent the corresponding number? What is the logarithm of a number?

217. To what is the sum of the logarithms of any two numbers equal? To what, then, will the addition of logarithms correspond?

*The sum of the logarithms of any two numbers is equal to the logarithm of their product.*

Therefore, *the addition of logarithms corresponds to the multiplication of their numbers.*

**218.** If we divide Equation ( 1 ) by Equation ( 2 ), member by member, we have,

$$\frac{10^m}{10^n} = 10^{n-n} = \frac{M}{N};$$

but since 10 is the base of the system, $m - n$ is the logarithm of $\frac{M}{N}$; hence,

*If one number be divided by another, the logarithm of the quotient will be equal to the logarithm of the dividend, diminished by that of the divisor.*

Therefore, *the subtraction of logarithms corresponds to the division of their numbers.*

**219.** Let us examine further the equations,

$$10^0 = 1$$
$$10^1 = 10$$
$$10^2 = 100$$
$$10^3 = 1000$$
$$\&c., \quad \&c.$$

It is plain that the logarithm of 1 is 0, and that the logarithm of any number between 1 and 10, is greater than

---

218. If one number be divided by another, what will the logarithm of the quotient be equal to? To what, then, will the subtraction of logarithms correspond?

219. What is the logarithm of 1? Between what limits are the logarithms of all numbers between 1 and 10? How are they generally expressed?

0 and less than 1. The logarithm is generally expressed by decimal fractions; thus,

$$\log 2 = 0.301030.$$

The logarithm of any number greater than 10 and less than 100, is greater than 1 and less than 2, and is expressed by 1 and a decimal fraction; thus,

$$\log 50 = 1.698970.$$

The part of the logarithm which stands at the left of the decimal point, is called the *characteristic* of the logarithm. The characteristic is always *one less than the number of places of figures in the number whose logarithm is taken.*

Thus, in the first case, for numbers between 1 and 10, there is but one place of figures, and the characteristic is 0. For numbers between 10 and 100, there are two places of figures, and the characteristic is 1; and similarly for other numbers.

### TABLE OF LOGARITHMS.

**220.** A table of logarithms is a table in which are written the logarithms of all numbers between 1 and some other given number. A table showing the logarithms of the numbers between 1 and 100 is annexed. The numbers are written in the column designated by the letter N, and the logarithms in the column designated by Log.

---

How is it with the logarithms of numbers between 10 and 100? What is that part of the logarithm called which stands at the left of the characteristic? What is the value of the characteristic?

220. What is a table of logarithms? Explain the manner of finding the logarithms of numbers between 1 and 100?

13*

TABLE.

| N. | Log. | N. | Log. | N. | Log. | N. | Log. |
|----|------|----|------|----|------|----|------|
| 1 | 0.000000 | 26 | 1.414973 | 51 | 1.707570 | 76 | 1.880814 |
| 2 | 0.301030 | 27 | 1.431364 | 52 | 1.716003 | 77 | 1.886491 |
| 3 | 0.477121 | 28 | 1.447158 | 53 | 1.724276 | 78 | 1.892095 |
| 4 | 0.602060 | 29 | 1.462398 | 54 | 1.732394 | 79 | 1.897627 |
| 5 | 0.698970 | 30 | 1.477121 | 55 | 1.740363 | 80 | 1.903090 |
| 6 | 0.778151 | 31 | 1.491362 | 56 | 1.748188 | 81 | 1.908485 |
| 7 | 0.845098 | 32 | 1.505150 | 57 | 1.755875 | 82 | 1.913814 |
| 8 | 0.903090 | 33 | 1.518514 | 58 | 1.763428 | 83 | 1.919078 |
| 9 | 0.954243 | 34 | 1.531479 | 59 | 1.770852 | 84 | 1.924279 |
| 10 | 1.000000 | 35 | 1.544068 | 60 | 1.778151 | 85 | 1.929419 |
| 11 | 1.041393 | 36 | 1.556303 | 61 | 1.785330 | 86 | 1.934498 |
| 12 | 1.079181 | 37 | 1.568202 | 62 | 1.792392 | 87 | 1.939519 |
| 13 | 1.113943 | 38 | 1.579784 | 63 | 1.799341 | 88 | 1.944483 |
| 14 | 1.146128 | 39 | 1.591065 | 64 | 1.806180 | 89 | 1.949390 |
| 15 | 1.176091 | 40 | 1.602060 | 65 | 1.812913 | 90 | 1.954243 |
| 16 | 1.204120 | 41 | 1.621784 | 66 | 1.819544 | 91 | 1.959041 |
| 17 | 1.230449 | 42 | 1.623249 | 67 | 1.826075 | 92 | 1.963788 |
| 18 | 1.255273 | 43 | 1.633468 | 68 | 1.832509 | 93 | 1.968483 |
| 19 | 1.278754 | 44 | 1.643453 | 69 | 1.838849 | 94 | 1.973128 |
| 20 | 1.301030 | 45 | 1.653213 | 70 | 1.845098 | 95 | 1.977724 |
| 21 | 1.322219 | 46 | 1.662758 | 71 | 1.851258 | 96 | 1.982271 |
| 22 | 1.342423 | 47 | 1.672098 | 72 | 1.857333 | 97 | 1.986772 |
| 23 | 1.361728 | 48 | 1.681241 | 73 | 1.863323 | 98 | 1.991226 |
| 24 | 1.380211 | 49 | 1.690196 | 74 | 1.869232 | 99 | 1.995635 |
| 25 | 1.397940 | 50 | 1.698970 | 75 | 1.875061 | 100 | 2.000000 |

EXAMPLES.

1. Let it be required to multiply 8 by 9, by means of logarithms. We have seen, Art. 216, that the sum of the logarithms is equal to the logarithm of the product. Therefore, find the logarithm of 8 from the table, which is 0.903090, and then the logarithm of 9, which is 0.954243; and their sum, which is 1.857333, will be the logarithm of the product. In searching along in the table, we find that 72 stands opposite this logarithm; hence, 72 is the product of 8 by 9.

2. What is the product of 7 by 12?

| | |
|---|---|
| Logarithm of 7 is, . . . . | 0.845098 |
| Logarithm of 12 is, . . . . | 1.079181 |
| Logarithm of their product, . . | 1.924279 |

and the corresponding number is 84.

3. What is the product of 9 by 11?

| | |
|---|---|
| Logarithm of 9 is, . . . . | 0.954243 |
| Logarithm of 11 is, . . . . | 1.041393 |
| Logarithm of their product, . . | 1.995636 |

and the corresponding number is 99.

4. Let it be required to divide 84 by 3. We have seen in Art. 218, that the subtraction of Logarithms corresponds to the division of their numbers. Hence, if we find the logarithm of 84, and then subtract from it the logarithm of 3, the remainder will be the logarithm of the quotient.

| | |
|---|---|
| The logarithm of 84 is, . . . | 1.924279 |
| The logarithm of 3 is, . . . | 0.477121 |
| Their difference is, . . . . | 1.447158 |

and the corresponding number is 28.

5. What is the product of 6 by 7?

| | |
|---|---|
| Logarithm of 6 is, . . . . | 0.778151 |
| Logarithm of 7 is, . . . . | 0.845098 |
| Their sum is, . . . . . | 1.623249 |

and the corresponding number of the table, 42.

# RECOMMENDATIONS OF DAVIES' MATHEMATICS.

DAVIES' COURSE OF MATHEMATICS *are the prominent Text-Books in most of the Colleges of the United States,* and also in the various Schools and Academies throughout the Union.

YORK, PA., *Aug.* 28, 1858.

*Davies' Series of Mathematics* I deem the very best I ever saw. From a number of authors I selected it, after a careful perusal, as a course of study to be pursued by the Teachers attending the sessions of the York Co. Normal School—believing it also to be well adapted to the wants of the schools throughout our country. Already two hundred schools are supplied with DAVIES' valuable *Series of Arithmetics;* and I fully believe that in a very short time the Teachers of our country *en masse* will be engaged in imparting instruction through the medium of this new and easy method of analysis of numbers. A. R. BLAIR,
*Principal of York Co. Normal School.*

JACKSON UNION SCHOOL, MICHIGAN, *Sept.* 25, 1858.

MESSRS A. S. BARNES & Co.:—I take pleasure in adding my testimony in favor of *Davies' Series of Mathematics,* as published by you. We have used these works in this school for more than four years; and so well satisfied are we of their superiority over any other Series, that we neither contemplate making, nor desire to make, any change in that direction. Yours truly, E. L. RIPLEY.

NEW BRITAIN, *June* 12th, 1858.

MESSRS. A. S. BARNES & Co.:—I have examined *Davies' Series of Arithmetics* with some care. They appear well adapted for the different grades of schools for which they are designed. The language is clear and precise; each principle is thoroughly analyzed, and the whole so arranged as to facilitate the work of instruction. Having observed the satisfaction and success with which the different books have been used by eminent teachers, it gives me pleasure to commend them to others. DAVID N. CAMP, *Principal of Conn. State Normal School.*

I have long regarded *Davies' Series of Mathematical Text-Books* as far superior to any now before the public. We find them in every way adapted to the wants of the Normal School, and we use no other. A unity of system and method runs throughout the series, and constitutes one of its great excellences. Especially in the Arithmetics the author has earnestly endeavored to supply the wants of our Common and Union Schools: and his success is complete and undeniable. I know of no Arithmetics which exhibit so clearly the philosophy of numbers, and at the same time lead the pupil surely on to readiness and practice. A. S. WELCH.

*From PROF. G. W. PLYMPTON, late of the State Normal School, N. Y.*

Out of a great number of Arithmetics that I have examined during the past year, I find none that will compare with *Davies' Intellectual* and *Davies' Analytical and Practical Arithmetics,* in clearness of demonstration or philosophical arrangement. I shall with pleasure recommend the use of these two excellent works to those who go from our Institution to teach.

*From C. MAY, JR., School Commissioner, Keene, N. H.*

I have carefully examined *Davies' Series of Arithmetics,* and *Higher Mathematics,* and am prepared to say that I consider them far superior to any with which I am acquainted.

*From JOHN L. CAMPBELL, Professor of Mathematics, Natural Philosophy, and Astronomy, in Wabash College, Indiana.*

WABASH COLLEGE, *June* 22, 1858.

MESSRS. A. S. BARNES & Co.:—GENTLEMEN: Every text-book on Science properly consists of two parts—the *philosophical* and the *illustrative.* A proper combination of abstract reasoning and practical illustration is the chief excellence in Prof. Davies' Mathematical Works. I prefer his Arithmetics, Algebras, Geometry, and Trigonometry, to all others now in use, and cordially recommend them to all who desire the advancement of sound learning. Yours, very truly, JOHN L. CAMPBELL.

PROFESSORS MAHAN, BARTLETT, and CHURCH, of the United States Military Academy, West Point, say of *Davies' University Arithmetic:*—

"In the distinctness with which the various definitions are given, the clear and strictly mathematical demonstration of the rules, the convenient form and well-chosen matter of the tables, as well as in the complete and much-desired application of all to the business of the country, the *University Arithmetic* of Prof. Davies is superior to any other work of the kind with which we are acquainted."

# RECOMMENDATIONS

OF

# PARKER & WATSON'S READERS

—————◆◆—————

*From* PROF. FREDERICK S. JEWELL, *of the New York State Normal School*

It gives me pleasure to find in the National Series of School Readers ample room for commendation. From a brief examination of them, I am led to believe that we have none equal to them. I hope they will prove as popular as they are excellent

*From* HON. THEODORE FRELINGHUYSEN, *President of Rutgers' College, N. J.*

A cursory examination leads me to the conclusion that the system contained in these volumes deserves the patronage of our schools, and I have no doubt that it will become extensively used in the education of children and youth.

*From* N. A. HAMILTON, *President of Teachers' Union, Whitewater, Wis.*

The National Readers and Speller I have examined, and carefully compared with others, and must pronounce them decidedly superior, in respect to literary merit, style, and price. The gradation is more complete, and the series much more desirable for use in our schools than Sanders' or McGuffey's.

*From* PROF. T. F. THICKSTUN, *Principal of Academy and Normal School, Meadville, Pa.*

I am much pleased with the National Series of Readers after having canvassed their merits pretty thoroughly. The first of the series especially pleases me, because it affords the means of teaching the "*word-method*" in an appropriate and natural manner. They all are progressive, the rules of elocution are stated with clearness, and the selection of pieces is such as to please at the same time that they instruct.

*From* J. W. SCHERMERHORN, A. B., *Principal Coll. Institute, Middletown, N. J*

I consider them emphatically *the* Readers of the present day, and I believe that their intrinsic merits will insure for them a full measure of popularity.

*From* PETER ROUGET, *Principal Public School No. 10, Brooklyn.*

It gives me great pleasure to be able to bear my unqualified testimony to the excellence of the National Series of Readers, by PARKER and WATSON. The gradation of the books of the series is very fine; we have reading in its elements and in its highest style. The fine taste displayed in the selections and in the collocation of the pieces deserves much praise. A distinguishing feature of the series is the variety of the subject-matter and of the style. The practical teacher knows the value of this characteristic for the development of the voice. The authors seem to have kept constantly in view the fact that a reading-book is designed for children, and therefore they have succeeded in forming a very interesting and improving collection of reading-matter, highly adapted to the wants and purposes of the school-room. In short, I look upon the National Series of Readers as a great success.

*From* A. P. HARRINGTON, *Principal of Union School, Marathon, N. Y.*

These Readers, in my opinion, are the best I have ever examined. The rhetorical exercises, in particular, are superior to any thing of the kind I have ever seen. I have had better success with my reading classes since I commenced training them on these than I ever met with before. The marked vowels in the reading exercises convey to the reader's mind at once the astonishing fact that he has been accustomed to mispronounce more than one-third of the words of the English language.

*From* CHARLES S. HALSEY, *Principal Collegiate Institute, Newton, N. J.*

In the simplicity and clearness with which the principles are stated, in the appropriateness of the selections for reading, and in the happy adaptation of the different parts of the series to each other, these works are superior to any other text books on the subject which I have examined.

*From* WILLIAM TRAVIS, *Principal of Union School, Flint, Mich.*

I have examined the National Series of Readers, and am delighted to find it so far in advance of most other series now in use, and so well adapted to the wants of the Public Schools. It is unequaled in the skillful arrangement of the material used, beautiful typography, and the general neat and inviting appearance of its several books. I predict for it a cordial welcome and a general introduction by many of our most enterprising teachers.

# RECOMMENDATIONS

## OF

# CLARK'S ENGLISH GRAMMAR.

We cannot better set forth the merits of this work than by quoting a part of a communication from Prof. F. S. Jewell, of the New York State Normal School, in which school this Grammar is now used as the text book on this subject:—

"Clark's System of Grammar is worthy of the marked attention of the friends of education. Its points of excellence are of the most decided character, and will not soon be surpassed. Among them are—

1st. "The justness of its ground principle of classification. There is no simple, philosophical, and practical classification of the elements of language, other than that built on their use or office. Our tendencies hitherto to follow the analogies of the classical languages, and classify extensively according to forms, have been mischievous and absurd. It is time we corrected them.

2d. "Its thorough and yet simple and transparent analysis of the elements of the language according to its ground principle. Without such an analysis, no broad and comprehensive view of the structure and power of the language can be attained. The absence of this analysis has hitherto precipitated the study of Grammar upon a surface of dry details and bare authorities, and useless technicalities.

3d. "Its happy method of illustrating the relations of elements by diagrams. These, however uncouth they may appear to the novice, are really simple and philosophical. Of their utility there can be no question. It is supported by the usage of other sciences, and has been demonstrated by experience in this.

4th. "The tendency of the system, when rightly taught and faithfully carried out, to cultivate habits of nice discrimination and close reasoning, together with skill in illustrating truth. In this it is not excelled by any, unless it be the mathematical sciences, and even there it has this advantage, that it deals with elements more within the present grasp of the intellect. On this point I speak advisedly.

5th. "The system is thoroughly progressive and practical, and as such, American in its character. It does not adhere to old usages, merely because they are venerably musty; and yet it does not discard things merely because they are old, or are in unimportant minutiæ not prudishly perfect. It does not overlook details and technicalities, nor does it allow them to interfere with plain philosophy or practical utility.

"Let any clear-headed, independent-minded teacher master the system, and then give it a fair trial, and there will be no doubt as to his testimony."

*A Testimonial from the Principals of the Public Schools of Rochester, N. Y.*

We regard Clark's Grammar as the clearest in its analysis, the most natural and logical in its arrangement, the most concise and accurate in its definitions, the most systematic in design, and the best adapted to the use of schools of any Grammar with which we are acquainted.

| | |
|---|---|
| C. C. MESERVE, | WM. C. FEGLES. |
| M. D. ROWLEY, | OHN ATWATER, |
| C. R. BURRICK, | EDWARD WEBSTER, |
| J. R. VOSBURG. | S. W. STARKWEATHER, |
| E. R. ARMSTRONG | PHILIP CURTISS. |

Lawrence Institute, *Brooklyn, Jan. 15, 1859.*

Messrs. A. S. Barnes & Co:—Having used Clark's New Grammar since its publication, I do most unhesitatingly recommend it as a work of superior merit. By the use of no other work, and I have used several, have I been enabled to advance my pupils so rapidly and thoroughly.

The author has, by an Etymological Chart and a system of Diagrams, made Grammar the study that it ought to be, interesting as well as useful.

MARGARET S. LAWRENCE, *Principal.*

---

## WELCH'S ENGLISH SENTENCE.

*From* Prof. J. R. Boise, A. M., *Professor of the Latin and Greek Languages and Literature in the University of Michigan.*

This work belongs to a new era in the grammatical study of our own language. We hazard nothing, in expressing the opinion, that for severe, searching, and exhaustive analysis, the work of Professor Welch is second to none. His book is not intended for beginners, but only for advanced students, and by such only it will be understood and appreciated.

# MONTEITH AND McNALLY'S GEOGRAPHIES

## THE MOST SUCCESSFUL SERIES EVER ISSUED.

## RECOMMENDATIONS.

A. B. CLARK, Principal of one of the largest Public Schools in Brooklyn, says:—
"I have used over a thousand copies of Monteith's Manual of Geography since its
adoption by the Board of Education, and am prepared to say it is the best work for
Junior and intermediate classes in our schools I have ever seen."

*The Series, in whole or in part, has been adopted in the*

New York State Normal School.
New York City Normal School.
New Jersey State Normal School.
Kentucky State Normal School.
Indiana State Normal School.
Ohio State Normal School.
Michigan State Normal School.
York County (Pa.) Normal School.
Brooklyn Polytechnic Institute.
Cleveland Female Seminary.
Public Schools of Milwaukie.
Public Schools of Pittsburgh.
Public Schools of Lancaster, Pa.
Public Schools of New Orleans.

Public Schools of New York.
Public Schools of Brooklyn, L. I.
Public Schools of New Haven.
Public Schools of Toledo, Ohio.
Public Schools of Norwalk, Conn.
Public Schools of Richmond, Va.
Public Schools of Madison, Wis.
Public Schools of Indianapolis.
Public Schools of Springfield, Mass.
Public Schools of Columbus, Ohio.
Public Schools of Hartford, Conn.
Public Schools of Cleveland, Ohio.
And other places too numerous to
mention.

They have also been recommended by the State Superintendents of ILLINOIS,
INDIANA, WISCONSIN, MISSOURI, NORTH CAROLINA, ALABAMA, and by numerous
Teachers' Associations and Institutes throughout the country, and are in successful
use in a multitude of Public and Private Schools throughout the United States.

*From* PROF. WM. F. PHELPS, A. M., *Principal of the New Jersey State
Normal School.*

TRENTON, *June* 17, 1858.

MESSRS. A. S. BARNES & Co.:—GENTLEMEN: It gives me much pleasure to state
that McNally's Geography has been used in this Institution from its organization in
1855, with great acceptance. The author of this work has avoided on one hand the
extreme of being too meager, and on the other of going too much into detail, while
he has presented, in a clear and concise manner, all those leading facts of Descriptive
Geography which it is important for the young to know. The maps are accurate and
well executed, the type clear, and indeed the entire work is a decided success. I most
cheerfully commend it to the profession throughout the country.
Very truly yours, WM. F. PHELPS.

*From* W. V. DAVIS, *Principal of High School, Lancaster, Pa.*

LANCASTER, PA., *June* 26, 1858.

DEAR SIRS:—I have examined your *National Geographical Series* with much
care, and find them most excellent works of their kind. They have been used in the
various Public Schools of this city, ever since their publication, with great success and
satisfaction to both pupil and teacher. All the Geographies embraced in your series
are well adapted to school purposes, and admirably calculated to impart to the pupil,
in a very attractive manner, a complete knowledge of a science, annually becoming
more useful and important. Their maps, illustrations, and typography, are unsur-
passed. One peculiar feature of McNally's Geography—and which will recommend
it at once to every practical teacher—is the arrangement of its maps and lessons;
each map fronts the particular lesson which it is designed to illustrate—thus enabling
the scholar to prepare his task without that constant turning over of leaves, or reference to a separate book, as is necessary with most other Geographies. Yours, &c.
Messrs. A. S. BARNES & Co., New York. V. W. DAVIS.

*From* CHARLES BARNES, *late President State Teachers' Association, and Superin-
tendent of the Public Schools at New Albany, Indiana.*

MESSRS. A. S. BARNES & Co.:—DEAR SIRS: I have examined with considerable
care the Series of Geographies published by you, and have no hesitation in saying
that it is altogether the best with which I am acquainted. A trial of more than a
year in the Public Schools of this city has demonstrated that *Cornell* is utterly unfit
for the school-room. Yours, &c. C. BARNES.

# RECOMMENDATIONS

## OF

# MONTEITH'S HISTORY OF THE UNITED STATES.

This volume is designed for youth, and we think the author has been unusual y successful in its arrangement and entire preparation. Books of the same design a too often beyond the full understanding of the scholar. As history is so much ne<sub></sub> rected in all our schools, the publication of such a work as this should be hailed with pleasure; for if scholars find their first studies of history pleasant, it will become a pleasure rather than a task. This is a book of 88 pages, and finely illustrated. It is in every way worthy of a place in every Public School in the State.—*Maine Teacher.*

This is a most capital work : just the thing for children. Our boy commenced the study of it the day it came to hand. It is arranged in the catechetical form, and is finely illustrated with maps, with special reference to the matter discussed in the text. It begins with the first discoveries of America, and comes down to the laying of the Atlantic Telegraph Cable. Many spirited engravings are given to illustrate the work. It also contains brief Biographies of all prominent men who have identified themselves with the history of this country. It is the best work of the kind we have seen.—*Chester County Times.*

## WILLARD'S HISTORIES.

*From* REV. HOWARD MALCOLM, D. D., *President of the University of Lewisburg.*

I have examined, during the thirteen years that I have had charge of a College, many School Histories of the United States, and have found none, on the whole, so proper for a text-book as that of Mrs. Willard. It is neither too short nor too long, all the space given to periods, events, and persons, is happily proportioned to their importance. The style is attractive and lucid, and the narrative so woven, as both to sustain the interest and aid the memory of the student. Candor, impartiality, and accuracy, are conspicuous throughout. I think no teacher intending to commence a history class will be disappointed in adopting this book.

MRS. L. H. SIGOURNEY, *the distinguished Authoress,* writes:

Mrs. Willard should be considered as a benefactress not only by her own sex, of whom she became in early years a prominent and permanent educator, but by the country at large, to whose good she has dedicated the gathered learning and faithful labor of life's later periods. The truths that she has recorded, and the principles that she has impressed, will win, from a future race, gratitude that cannot grow old, and a garland that will never fade.

DANIEL WEBSTER *wrote, in a letter to the Author:*

I cannot better express my sense of the value of your *History of the United States,* than by saying I keep it near me as a book of reference, accurate in facts and dates.

## DWIGHT'S MYTHOLOGY.

The mythology of the Grecians and Romans is so closely interlinked with the history and literature of the world, that some knowledge of it is indispensable to any scholarly familiarity with either that history or literature. We have seen no book so convenient in size that contains so full and elegant an exposition of mythology as the one before us. It will be found at once a most interesting and a most useful book to any one who wishes an acquaintance with the splendid myths and fables with which the great masters of ancient learning amused their leisure and cheated their faith - *Michigan Journal of Education.*